W9-CMG-735

CLARENDON LIBRARY OF LOGIC AND PHILOSOPHY

General Editor: L. Jonathan Cohen

TRUTH PROBABILITY AND PARADOX

TRUTH PROBABILITY AND PARADOX

Studies in Philosophical Logic

J. L. MACKIE

OXFORD
AT THE CLARENDON PRESS
1973

Oxford University Press, Ely House, London W. 1

GLASGOW NEW YORK TORONTO MELBOURNE WELLINGTON
CAPE TOWN IBADAN NAIROBI DAR ES SALAAM LUSAKA ADDIS ABABA
DELHI BOMBAY CALCUTTA MADRAS KARACHI LAHORE DACCA
KUALA LUMPUR SINGAPORE HONG KONG TOKYO

Printed in Great Britain
at the University Press, Oxford
by Vivian Ridler
Printer to the University

To my teachers in philosophy

JOHN ANDERSON

and

W. G. MACLAGAN

'For words are wise mens counters, they do but reckon by them: but they are the mony of fooles . . .'

HOBBES, *Leviathan*

PREFACE

PHILOSOPHY, to be any good, must be analytic; but conceptual analysis is not the whole of philosophy. Any genuine progress with philosophical problems requires the sort of argument that takes account of alternative possibilities, that formulates suggestions precisely enough to allow them to be fairly examined and tested, that pays attention to the meanings of the words it uses, and that reflects critically on its own procedures. But the aim is to make progress with substantive questions, to apply our concepts to reality or to consider how far they are applicable, not merely to analyse or clarify those concepts themselves.

This book is concerned mainly but not exclusively with analytical tasks. Conditional statements play a large part in all our thinking and enter into the analysis of many philosophically important concepts. Chapter 3 discusses how they should be understood and argues that inadequate understanding of them has led to mistaken views about causation and natural law. We use the concepts of dispositions, dispositional properties, and powers extensively in describing both ourselves and the physical world; Chapter 4 shows how these concepts may be clarified (partly with the help of conditionals) but also considers to what sorts of real entities our dispositional descriptions point. The concept (or rather cluster of concepts) of probability is important in several ways: it comes in wherever there are good but inconclusive reasons for believing anything; it enters into the explanation of fundamental physical processes; and some grasp of it is a prerequisite for discussion of the problems of scientific knowledge and induction. But this topic is confused not only by the clash of rival theories and the real multiplicity of concepts in use but also by the growth of technical and mathematical elaboration. Chapter 5 attempts to sort out this confusion in a non-technical way, and uses the previous chapter's treatment of dispositions to throw light on the propensity view of probability and the notion of objective chance. Chapters 2 and 6 are closely related: the simple, classical, concept of truth, which is central in our understanding of the world, is defended in Chapter 2 and contrasted with various rival theories. But it is challenged by certain logical paradoxes, whose solution seems to call for distinctions that undermine it and that lead to a rival theory. Chapter 6 argues, in reply, that the only fully satisfactory solution of the paradoxes is one which uses the concept of truth defended in Chapter 2, criticizing, on the way, a number of widely held but (I believe) mistaken views about the paradoxes. Chapter 1 distinguishes different kinds of analysis and shows how the 'paradox of analysis' can be avoided in several ways; its general theses are illustrated by many parts of the later chapters.

Though there are these links between them, each chapter is a fairly separate study of a topic in philosophical logic. They are unified by two controlling ideas. One is that these topics can be illuminated, and the problems that arise within them resolved, by the coherent application of fairly simple, common-sense, perhaps old-fashioned ways of thinking. I find myself defending a view of truth that goes back to Aristotle and a solution of the paradoxes that was sketched by Carneades. These topics do not need to be so difficult or so technical as they are sometimes made to appear. It is as well that they do not, for all the concepts discussed are of instrumental importance: we need the notions of analysis, truth, conditionals, dispositions, probability, and contradiction for many philosophical purposes as well as for some everyday ones: we cannot afford to let them become the preserve of specialists. The ordinary concept of truth would lose its force if it were mixed up with the structural analysis of a language. Probability is not only the guide of life, it is also, in almost every field, the guide of thought, and we cannot abandon it to the mathematicians, statisticians, and physicists. The other controlling idea is that we need to take precautions against being fooled, in more than one possible way, by the words (and the symbols) we employ, and in particular that informal critical reflection is needed as a corrective to formal and mechanical procedures.

These two ideas come together. The more a topic becomes entangled in technical detail, and the more professional capital is invested in its elaboration, the harder it is to bring critical reflection to bear upon the points where it is needed. But the harder this task is, the more essential it becomes. I have therefore tried to follow the advice of Empedocles to 'consider everything in the way it is clear', though I cannot be sure that I have succeeded. Of course, I cannot pretend to have dispensed altogether with technical apparatus and jargon; those who try to do so usually succeed only in inventing new jargons of their own. I would hope rather that a small protective inoculation with useful techniques may give us the strength to resist more damaging complications.

Chapter 2 incorporates but adds extensively to an article, 'Simple Truth', published in the *Philosophical Quarterly*, 20, no. 81, October 1970, which is reprinted by permission of the editor. Some parts of Chapter 3 reproduce parts of an article 'Counterfactuals and Causal Laws', published in *Analytical Philosophy* (First Series) edited by R. J. Butler—though my views about conditionals have changed in some fundamental respects since that article was written—and this material is used by permission of Basil Blackwell and Mott Limited. I have to thank the editor and the publishers in question.

I have also to thank the editor of *Critica* for permission to use, in Chapter 6, material which was included in an article, 'What can we learn from the paradoxes?', in that periodical, and the publishers of *Analysis* for permission to reprint, in the Appendix, the late Thomas Storer's note on 'Miniac', from *Analysis*, 22 (1961–2).

I have also to thank the Radcliffe Trustees for giving me a Radcliffe Fellowship, the prospect of which encouraged me to begin this work and the enjoyment of which has enabled me to complete it; also the Master and Fellows of University College for allowing me to accept this; and Mr. Peter Singer who, as Radcliffe Lecturer, has been doing the undergraduate teaching which would otherwise have been my responsibility.

I would also like to thank Mrs. M. Boldero and Mrs. E. Hinkes for typing the book.

To many people with whom I have discussed these topics I am indebted both for the original stimulus to develop the views expressed here and for criticisms of earlier formulations. I would mention particularly Mr. M. G. Evans and Mr. J. H. McDowell in connection with Dr. Chapter 2, D. H. Mellor with Chapter 4, and Mr. D. C. Stove and Professor R. G.

Swinburne with Chapter 5. But I have adhered obstinately on a number of issues to views which they would reject. I am very happy to find that Professor W. C. Kneale, in articles published in *Mind* for April 1972 and in the *British Journal for the Philosophy of Science* for November 1971 and August 1972, has reached, with regard to truth and the paradoxes, conclusions very close to those which I reach in Chapters 2 and 6, though by a somewhat different route. But I did not read these articles until after I had completed my own account.

J. L. M.

February, 1972

CONTENTS

1

POSSIBILITIES OF ANALYSIS

A CHARGE commonly brought against analytical philosophy is that it is concerned solely with the meanings of words and that this is a trivial and unexciting pursuit, which contrasts disadvantageously with more enterprising, constructive, and speculative kinds of philosophy. The best and only complete reply to this charge is to give examples of work in analytical philosophy which is not concerned solely with the meanings of words and which makes serious, but at the same time carefully argued, contributions to our understanding of the world, of our knowledge of the world, of human beings and human actions. And of course it is not difficult to do this. But it may be worth while also to consider in general some of the ways in which analysis (whether linguistic or not) can escape triviality.

We may start by considering the 'paradox of analysis', posed as follows by Langford as a difficulty for Moore's view of philosophy:

> Let us call what is to be analyzed the analysandum, and let us call that which does the analyzing the analysans. The analysis then states an appropriate relation of equivalence between the analysandum and the analysans. And the paradox of analysis is to the effect that, if the verbal expression representing the analysandum has the same meaning as the verbal expression representing the analysans, the analysis states a bare identity and is trivial; but if the two verbal expressions do not have the same meaning, the analysis is incorrect.[1]

Several things are worth noting about this formulation. First, it is assumed that what is to be analysed is not a word or verbal expression, but something represented by a verbal expression: Moore himself suggests that the only thing that could properly

[1] C. H. Langford, 'The Notion of Analysis in Moore's Philosophy', *The Philosophy of G. E. Moore*, ed. P. A. Schilpp, pp. 321–42, esp. p. 323.

be called analysing a verbal expression is to say what words or letters it contains and in what order,[2] and of course that is not what we want. Secondly, there is then a temptation, to which Moore (though not Langford) succumbed, to say that what is analysed is a concept, and that the analysans must be 'in some sense' the same concept; and then the analysis does seem to collapse into 'a bare identity'. Thirdly, it is assumed that the analysis will take some such form as 'An *X* is a *Y*' or 'To be an *X* is to be a *Y*', where '*X*' and '*Y*' are the verbal expressions representing the analysandum and the analysans respectively.

I shall show that there is more than one way in which an analysis can escape this paradox, and that some, though not all, of these leave room for analyses that are of philosophical interest.

It will make things clearer if we introduce three characters: *A*, the analyser, the person who offers the analysis; *B*, an ordinary competent user of the language and the concepts in question; and *C*, a foreigner, with only a partial knowledge of the language that is being used.

First, suppose that '*X*' and '*Y*' are two verbal expressions which are synonymous (or as nearly synonymous as matters for the purpose in hand) in the language *L*, but *C* knows the meaning of '*Y*' in *L* but not the meaning of '*X*' in *L*, while *A* knows that *C*'s knowledge is thus limited, and *C* knows that *A* knows this: they are both in the clear about their relational situation with respect to *L*. Then if *A* were to say to *C*: 'In *L*, "*X*" means *Y*', he would be saying something which was true, non-trivial, and informative to *C* though not of course informative to *A* himself or, say, to *B*. What is more curious is that *A* could convey this same information to *C* by saying simply 'An *X* is a *Y*' or 'To be an *X* is to be a *Y*'. For example, instead of saying 'In English, "oculist" means eye-doctor' he could say 'An oculist is an eye-doctor' or 'To be an oculist is to be an eye-doctor'. How does this come about? How can we convey information about meaning without mentioning meaning? It is not hard to explain this. Since *C* does not yet know what 'oculist' means, when he hears *A* say 'An oculist . . .' he can only take this as equivalent to 'Whatever is called an oculist in English . . .'. Now if *A* had said to *B*, for example, 'An oculist is a menace', and *C* had merely overheard

[2] 'A Reply to my Critics', *The Philosophy of G. E. Moore*, p. 665.

this, *C* could take this only as saying that whatever is called an oculist in English is a menace, and (even if he understood 'a menace') he would not have learned much, since the subject-phrase would have failed to pick out, for him, any definite subject. But since, by hypothesis, *A* had said 'An oculist is an eye-doctor' *to C*, he cannot have been using the predicate-expression 'an eye-doctor' analogously with 'a menace' in the sentence addressed to *B*; he must have intended *C* to be able to get some definite information out of his remark; he must therefore have been using 'an eye-doctor' not as a further description of whatever is called an oculist in English but as a way of specifying what this is; and *C*, knowing that *A* is aware of *C*'s limitations, is able to take *A*'s remark in the way intended. A similar explanation applies to the form 'To be an oculist is to be an eye-doctor'. Spoken by *A* to *C* with the mutual knowledge we have assumed, this could not be analogous to, say, 'To be a housewife is to be a slave': the latter is used to assert the further description 'slaves' of any persons who satisfy the description 'housewives', whereas *C* cannot yet use the description 'oculists', and *A* knows this. The only interpretation consistent with what *A* and *C* know about each other is that which makes this formula convey precisely the information that 'oculist', in English, means eye-doctor.[3]

What if *A* uses one of these sentences, say 'An oculist is an eye-doctor', to *B*? We then have a choice between two unattractive interpretations. Either *A* is simply using the word 'oculist' with its standard meaning and stating a sheer tautology, just as if he had said 'An eye-doctor is an eye-doctor', or he is rather insultingly treating *B* as if he were in *C*'s position, and telling *B* what he already knows, that 'oculist' means eye-doctor. The second interpretation is the more natural: to repeat a known truth is not quite as conversationally objectionable as to state a simple tautology.

It is sometimes supposed that a definition is a tautology. But this is clearly wrong. Several kinds of statements are called definitions, but even the two which qualify best for being called verbal definitions are far from tautologous. One of these is the

[3] It will be obvious that this explanation makes use of ideas about meaning developed by H. P. Grice, e.g. in 'Meaning', *Philosophical Review*, 66 (1957), 377–88, and in his (unpublished) William James Lectures.

lexical definition, which reports how a word is in fact used, in particular what sense or senses it has now or has had at various times; this is clearly informative about the language, it is or should be based on empirical research, and is in fact quite likely to be wrong: the dictionary may well tell us that a word was not used with a certain sense before 1850 when in fact it was so used a century earlier. The other sort of verbal definition is the stipulative definition, a proposal to use a word in a certain way in a certain piece of work; being a proposal it cannot be true or false, and *a fortiori* cannot be tautologous. Of course the corresponding lexical definition will be true (but still not tautologous) of a piece of speech or writing which conforms to such a stipulation, or of that speech or writing interpreted as so conforming. But what may mislead us here is the fact that such a form of expression as 'An *X* is a *Y*' has at least two and possibly three uses. As we have seen, it can be used to convey just that information which would be conveyed by our explicit lexical definition 'In such-and-such a body of speech, "*X*" means *Y*'. With some strain, it could be used to express the corresponding stipulative definition. But, thirdly, the speaker could be just *using* the terms '*X*' and '*Y*', and not even obliquely conveying any information or stipulation about either of these terms; and if he uses '*X*' in a way of which the previous lexical definition would have been true, or in conformity with the previous stipulative definition, *then* his remark 'An *X* is a *Y*' will be a tautology. The same form of words can be used either obliquely, so that '*X*' is defined by it, or straightforwardly, so that '*X*' is merely used in it; and if '*X*' is used in the sense which this form of words could have been used to define, this straightforward use will be a tautology. The same expression can be either a definition or a tautology; but it cannot be both at once.

We have found, then, one way in which an analysis of the form Langford indicates could be true, non-trivial, and informative. But this way of escaping the paradox allows an analysis to be informative only to those who have an incomplete grasp of the language used. The information conveyed is about the meaning of a certain expression. This is not what Moore called analysis of a verbal expression (it does not say simply that the word 'dog' consists of the letters 'd', 'o', 'g', in that order), but it is not what we could plausibly call analysis of a concept either. It gives the meaning of an expression by associating it with a concept which the recipient of the analysis already has. It would seem that an analysis of this kind would not be of philosophical interest: philosophical analysis is carried on among persons who are presumed already to have a grasp of the language they are using. Is there any way in which this account can be modified or developed so that an analysis could be informative even to such persons?

To have a certain concept may be to have at least a certain habit or power of recognition. Someone who has a concept of a tree is able to classify observed objects as trees or as not being trees, and it is a minimal requirement for his having the ordinary concept of a tree that his classifications in this respect should agree pretty well with those of most other people except in so far as they could be explained as mistakes due to optical distortions and the like. Someone who speaks another language may have the same concept of a tree as we have if he groups the same class of objects together under a term in his language; we might also say that someone (or some animal) that does not use a language at all, or who, if he uses a language, has no term in it that can be translated as 'tree', still has the same concept of a tree as we have if he gives other evidence of classifying the right set of objects together. But these are only minimal requirements, they do not constitute a sufficient condition for having a, or the, concept of a tree.

Where there is such a habit or power of recognition, the person who has it will be responding to certain (in this case perceptual) cues; he will be in some way following rules and using criteria; but he may not know to what cues he is responding, he may not be consciously obeying whatever rules he is following. Let us suppose that he is not. Then there will be room for an analysis to be informative by making someone aware of the rules he has been unconsciously following. For example, someone, say *B*, may have a concept of a cube such that his classification of objects as cubes coincides (within reasonable limits of approximation, and apart from mistakes which could be somehow explained) with the standard one. And the cues by which he is guided, without knowing that he is, may be that the objects should be solid, with four sides, a top, and a bottom, all of which are flat and square. Let us suppose that these are also the cues by which most people are guided in their classification of things as cubes. Then *A* may analyse the concept of a cube, which *B* already shares, by saying that a cube is a solid object with four sides, a top, and a bottom, all of which are flat and square, and this may be both a correct analysis of *B*'s concept and informative to *B*.

That its substance can be both correct and informative is not now a puzzle; the statement 'You, in classifying things as cubes, use the following cues . . .' can clearly be both true and informative just because someone may use cues without knowing what they are. But we might still ask, how does the explicit formulation 'An *X* is a *Y*' escape tautologousness for someone whose concept of an *X* is that of a *Y*? The answer in this case is analogous to that in the case of *C*. For *B*, the phrase 'a cube', used by *A*, is equivalent to 'Whatever you, *B*, call a cube'; it introduces the denotation of the term 'cube', the range of things to which *B*, in agreement with most other people, applies this term. But the predicate 'a solid object etc.' introduces the connotation which is indeed implicit in *B*'s way of determining the denotation, but of which he was not previously aware. The formulation 'An *X* is a *Y*' predicates of the denotation of the term '*X*' the connotation which was used, but not known to be used, in fixing that denotation. It is misleading to read it as saying 'The concept represented by "*X*" is identical with the concept represented by "*Y*"'. But even this can be construed so as to be at once true and informative, if we allow for some equivocation in the uses of the word 'represented'. The habit of recognition *exercised* in *B*'s use of the word 'cube' is *describable* as the taking account of a thing's being a solid object, etc. But the information does not, and could not, lie in a certain concept's being identical with itself, but in the fact that these two different descriptions apply to the same concept, i.e. habit of recognition. This, I think, is the truth that underlies Langford's plausible but obscure suggestions that the analysans is more articulate and less idiomatic than the analysandum.

This, then, is a second possible way of escaping the paradox of analysis. The particular example I have used is not itself of philosophical interest, but some philosophically interesting analyses are of this general kind.

I have used an example in which a term was used in response

to *perceptual* cues; but this is not essential: the use of a term may be controlled by its implicit logic or depth grammar. *B*, being a competent speaker of English, knows how to use the phrase 'the highest mountain in New Zealand', and, let us suppose, says assertively, 'Mount Egmont is the highest mountain in New Zealand'. He may not be explicitly aware of the fact that in using this phrase he is guided by general logical rules for phrases of the form 'the *X*-est *Y*' which, applied to this case, make 'Mount Egmont is the highest mountain in New Zealand' equivalent to 'Mount Egmont is a mountain in New Zealand and for all *z*, if *z* is a mountain in New Zealand and is not Mount Egmont, Mount Egmont is higher than *z*'. But he is so guided: as soon as he learns that Mount Cook is another mountain in New Zealand and that Mount Egmont is not higher than Mount Cook, he will give up his previous assertion. There is therefore room for an analysis which is true of *B*'s use of a phrase but informative to *B* where what it informs him about is not the perceptual cues but the implicit logic which governs his use of that phrase. A similarly informative analysis equates 'a small elephant' with 'an elephant which is smaller than most elephants'. This and similar examples reveal a class of adjectives which are not simple predicates but operators upon other predicates. What would be of great philosophical interest, if it were correct, is Geach's claim that 'good' is in a somewhat analogous way an operator on other predicates, and that some important philosophical theories of ethics have been based partly on the failure to take account of this fact.[4]

The paradox of analysis was a paradox only because there was a conflict between our various assumptions and requirements. We wanted the analysis to be of a concept, and to be new and informative to someone who already possessed and employed that concept; we wanted it not merely to teach the meaning of a word or phrase to someone who could not yet use it. But we also tended to assume that a concept was just whatever someone who employed it intended to record or convey by the use of the associated words. Together, these would make the paradox insoluble: since a person cannot be ignorant of what he himself intends to record or convey, a correct analysis

[4] P. T. Geach, 'Good and Evil', *Analysis*, 17 (1956), 33–42, reprinted in *Theories of Ethics*, ed. Philippa Foot, pp. 64–73.

of a concept in that sense could not be informative to someone who already employed it. Our (second) way of solving or escaping the paradox was to recognize two other aspects or elements of a concept: the perceptual cues involved in a habit or power of recognition and the implicit logic of a linguistic expression or form of construction. These may be, though of course they need not be, unknown to the person who employs the concept. But having rightly stressed these (in order to escape the paradox) we should not deny or ignore this other aspect of a concept, what the speaker intends to record or convey. This may coincide with either or both of the other aspects—the speaker may intend to convey, in calling something an *X*, that it has just those features of which he is taking account, perceptually or logically, in calling it an *X*—or they may partly coincide, or either may include but overlap the other.

These various aspects may all be important in the attempt to analyse, for example, the concept of causation. Part at least of Hume's way of tackling this problem, part of what he is doing when he looks for the impression from which the idea of necessary connexion is derived, is to try to find the perceptual cues which govern our recognition that this *A* caused this *B*. Another approach is to try to formulate the implicit logic of 'This *A* caused this *B*'. A first attempt might be 'If this *A* had not happened, this *B* would not'; we could go on to criticize this by constructing cases in which a speaker would be prepared to use 'This *A* caused this *B*' but not 'If this *A* had not happened, this *B* would not', or vice versa; and on the basis of any successful criticisms of this sort we might try to construct an improved logical analysis. Hume's first 'definition', which we can paraphrase as 'This *A* contiguously preceded this *B*, and every *A* contiguously precedes a *B*', is again a not very successful attempt at a logical analysis. But Hume thought that there is also something else which anyone who ordinarily uses causal language intends thereby to record or convey, namely that there is some sort of *tie* or *link* between cause and effect, so that the one as it were pushes or pulls the other into existence. Since this part of what one intends to record or convey must also be, for Hume, an 'idea', he is committed to finding an (internal) impression from which it is derived and to giving a psychological explanation of how this gets mixed up with the habit of recognition by which we identify cases of causation.

Similar complications contribute to other traditional philosophical problems, particularly in the theory of knowledge. The cues used in our habit of recognition for, say, anger in other people are behavioural—though the behaviour may be wholly or partly verbal. An analysis of the implicit logic of 'He is angry' would show this aspect of the concept to be at least partly dispositional—I shall be considering in Chapter 4 what this means. But part of what I intend to record or convey when I say 'He is angry' is that there is, in the object identified by the word 'He', something very like what I have experienced in myself from time to time and have learned (initially, no doubt, from other people's ascriptions of anger to me) to call being angry. We all know this, though philosophers with one or other thesis to defend may close their eyes to it. Any ordinary speaker who says that another person is angry intends to ascribe to that other person something more than could be observed from the outside, even when predictions and retrodictions and tendencies are included, and the something more is what one knows about being angry from the inside.

Analysis of such a concept as anger will take us so far; but beyond this point we are liable to become victims of one or other of two rival dogmas. One is the verificationist dogma. Some concepts contain only those features which provide perceptual cues for our habit of recognition. Others have implicit logical complexities, but, according to this dogma, not only are the ultimate components of any logical analysis associated with perceptually based habits of recognition, but also the logical construction itself must be such that any assertion that utilizes it remains verifiable at least indirectly and at least in principle. (Stronger but less plausible forms of verificationism dispense with these qualifications.) Consequently, I cannot really intend to convey that other human bodies contain something like the feelings of anger that I myself experience: if I think that I am asserting this, I have failed to understand the logic of the language I speak. The other dogma is that our ordinary language, ordinary concepts, and ordinary conceptual scheme are all in perfect order. Since our ordinary concept of anger does contain the various aspects mentioned in the previous paragraph, and since behavioural cues are logically adequate criteria for my ascription of anger to someone else, this concept of anger—or the conceptual scheme, including other persons and myself, to which it belongs—in itself justifies the glueing together of the behavioural criteria for anger in other persons and the introspectible feelings whose presence I intend to record. The concept glues them together, and the concept is in order.

But these are both dogmas, and equally unjustified. I have deliberately given each in a crude form, so that their lack of justification is the more obvious: but the principles are quite general and will apply to subtler versions as well. There is no good reason why the logic of our language should be subject to verificationist constraints, no good reason why we should not intend to record or convey something that cannot be verified even indirectly, even in principle. At least, there is nothing to stop us from doing so *meaningfully*, unless the qualification 'in principle' is taken so widely that every assertion with a declarative meaning constructed out of ultimately empirical components is on that account alone called verifiable in principle, so that the verificationist theory of factual meaning collapses into a constructive one. And there is nothing to stop us from doing so *reasonably*, unless the qualification 'indirectly' is taken so widely that anything that could constitute a reason for belief is called an indirect verification. On the other hand, there is no good reason for supposing that our ordinary concepts are in perfect order. There may, indeed, be systematic ways of defending our ordinary conceptual scheme. Strawson has contrasted descriptive with revisionary metaphysics, where the former describes the existing conceptual scheme, the latter seeks to change it.[5] But this contrast neglects a third possibility, conservative metaphysics, which is not content to describe the existing scheme, but tries to justify and defend it, and it is under this third heading that much of Strawson's own argument would fall. His method of defence contains, in effect, two stages. One stage is to show that the existing conceptual scheme hangs together, that its parts are mutually dependent, so that the sceptic cannot consistently challenge one item—say, other minds—while retaining the rest. The other stage is to show that

[5] P. F. Strawson, *Individuals*, pp. 9–11.

a scepticism so extreme that it questions the whole existing conceptual scheme just cannot be formulated or entertained. In fact, both stages of this defence can be challenged, but it is an *argument*, to be examined on its merits—which I cannot do here. What I want to reject now is not this but the sheer *assumption* that ordinary language is in order, that once analysis has separated the components of any ordinary concept, its very ordinariness will glue them together again. Merely to employ a concept may be to make an implicit assertion, for example to assert that what users of the associated terms ordinarily intend to record or convey goes along with the features on which the habit of recognition which is part of that concept relies. This assertion cannot be eliminated on the ground that we cannot mean anything beyond what is utilized by the habit of recognition; nor can it be accepted without further question on the ground that the concept is an ordinary one. In each case there will be an issue to be further examined and discussed.

I have suggested that an analysis of the (or a) concept of a cube, and an analysis of the concept of causing, may be (in broad outline at least) of the same form and may escape in roughly the same way the triviality threatened in the paradox of analysis. And yet the first of these is not, and the second is, of philosophical interest. Can we explain why? The explanation lies partly in the difference between the roles of the concepts of a cube and of causing. The former determines the meaning of one word, and little else. But 'cause' is not an isolated word, it is at least a member of a set of near-synonyms—'bring about', 'produce', 'determine', 'is responsible for', 'result in', and so on—and of course these have near-equivalents in many other languages. This in itself makes it more appropriate to say that we are studying a concept than that we are studying the meaning of the verb 'to cause'. But besides this, the concept of causing is not an isolated concept. If we attempt to give a logical analysis of knowing, remembering, observing, of many examples of doing something, of moral responsibility, and so on, it is at least a plausible suggestion in each case that some causing will have to be mentioned in the analysis.

But there is still another factor that adds to the philosophical importance of the analysis of causation, and, as it happens, introduces a fourth way in which the paradox of analysis may be avoided. Given that we have a fairly determinate way of recognizing sequences and processes as causal—whether this rests on the implicit logic of causal language or on perceptual cues or both—so that we can mark out a fairly well-defined class of occurrences of causation, of cases where this caused that, we can ask the question, 'Is there anything common to and distinctive of the members of this class?' There is no need for the answer to this question to be exhausted by an account of the perceptual cues and the implicit logic involved in the method by which we have selected this class of cases. It need not even include everything mentioned in the logical analysis: this might well include items which cannot be truly ascribed to the objective occurrences about which we use them. For example, the implicit logic of causal statements may well include contrary-to-fact conditionals, but such conditionals (as I shall argue in Chapter 3) cannot be taken as literally descriptive of any objective states of affairs. And certainly the answer to our question need not include everything that we ordinarily intend to record and convey when we use causal statements: Hume might be right in thinking that we intend to convey that there is some sort of link between cause and effect, and yet that we are systematically wrong, that there is in fact no such link as we suppose. There is, then, a problem about the *factual* analysis of causation; we have the task of finding out what goes on in the world in those sequences and processes that we mark off as causal.

Since this is factual rather than conceptual analysis, it has no difficulty at all in escaping the paradox. There is not the slightest reason for supposing that anyone who successfully employs the ordinary concept of causing must know all that is in fact common to and distinctive of those sequences to which he has learnt to apply causal language in the standard, conventional, way. Indeed, the danger here is rather that philosophers may say that just because this is a factual investigation, it can be no business of theirs. But there is a topic here to be investigated, and if philosophers do not study it, it is hard to see who will. It is too broad and general a topic for any of the particular sciences. Besides, it would be almost impossible to pursue this question without first, or at the same time, doing those kinds of conceptual analysis that we have already distinguished. The conceptual and the factual analyses of causation are entwined with one another, and this, I suggest, helps to add philosophical importance to the conceptual side of the inquiry.

It is not difficult to describe an analogous factual analysis of a cube. Anything that satisfies the analysis given above for *B*'s concept of a cube (a solid object with four sides, a top, and a bottom, all of which are flat and square) will also satisfy a variety of other descriptions: e.g. it will be a regular solid with twelve edges. This is not logically equivalent to the analysis offered, but that anything which satisfies the previous analysis also satisfies this description is a truth of Euclidean geometry, or perhaps rather of the practical drawing-board geometry which is an approximate (and, within the limits of that approximation, correct) interpretation of the Euclidean system. We could call this a factual analysis of a cube, parallel to the projected factual analysis of causation; but it clearly belongs to some sort of mathematics, not to philosophy. It therefore does not reflect any further philosophical interest upon the conceptual analysis of a cube, as the factual analysis of causation does upon the corresponding conceptual analysis.

Nor is causation unusual in this respect. Philosophers who are concerned with perception, or with memory, or with personal identity, to take just three central and well-worn topics, are committed not merely to conceptual analyses of the kinds we have distinguished but also to saying what goes on when someone perceives something, or remembers something—for it is this, rather than any purely conceptual analysis, that will show what, and how, perception and memory can contribute to knowledge—and what, if anything, is really identical about each item which we take as instantiating the concept of one and the same person. Philosophers, then, do naturally and almost inevitably engage in factual analysis. Admittedly the factual issues in question are very broad ones, they are liable to lead on to or involve the choice of this or that conceptual scheme, and such issues can be contrasted with particular factual questions. Two rival conceptual schemes might each try to accommodate, somehow, the same set of particular facts, and this tempts us to contrast 'factual questions'—that is, questions of particular fact—with questions of interpretation. But the question whether this or that conceptual scheme is appropriate, whether particular facts are to be interpreted in one general way or another, is itself one to which there is some true answer, however elusive and hard to determine it may be. A big fact is still a fact, even if it is also the answer to a problem of interpretation.[6]

If we grant that a philosophical account of, say, perception is a piece of factual rather than conceptual analysis, we must take note of some complications. It would appear that anyone who perceives something makes, or is in a position to make, certain judgements. These judgements will have some implicit logic, they will call for conceptual analysis. The factual analysis of perceiving will then have to *include* the conceptual analysis of the perceiver's judgements. But it would obviously be a mistake to treat the contents of this conceptual analysis as if they were in themselves an adequate factual analysis of perceiving: an account of the meaning of 'I see . . .' is not necessarily a correct description of seeing. But there is a considerable temptation for philosophers to substitute the one for the other.

Another complication is this. Useful as the contrast between words and the world may be in many contexts, it can at times become misleading, in that words, too, are part of the world. In recent years philosophers have been led, quite naturally and appropriately, to move from a study of the implicit logic of various forms of expression to a factual analysis of language in use. To take first a very minor example: it is a matter of the implicit logic of an expression that 'Neither James nor John is fat' is equivalent to 'James is not fat and John is not fat'. But if we explain why this is so we shall be led to formulate a fragment of the structural semantics[7] of the English language and perhaps to touch upon its transformational grammar. To take a more serious example: Strawson in 'Intention and Convention in Speech Acts' is clearly engaged in factual analysis of language.[8] He is concerned primarily not with Austin's concept of an illocutionary act, but with the question, 'What concepts of

[6] Contrast what is said by A. J. Ayer in 'Philosophy and Language', *The Concept of a Person*, pp. 1–35, esp. pp. 20–2.

[7] i.e. an account of the contribution which the internal structure of complex sentences of a language makes to their meaning. More is said about this in Chapter 2, § 3.

[8] *Philosophical Review*, 73 (1964), 439–60, reprinted in *The Philosophy of Language*, ed. J. R. Searle, pp. 23–38.

illocutionary acts and the like do we *need* if we are adequately to describe linguistic communication?' And many other philosophers have strayed into linguistics by similar paths.

I have not so far explicitly queried the suggestion, in Langford's way of formulating the paradox, that an analysis can always be summed up in some such form as 'An *X* is a *Y*'. Analyses of causation, perception, and so on would obviously require some development of this at least: they might start, for example, 'For an event *A* to cause an event *B* is . . .'. But what is far more important is that once we come to factual analysis of language and language use we shall often need a kind of account which cannot be squeezed into anything like the form 'An *X* is a *Y*', where '*X*' is the linguistic expression or construction under discussion. We may need to explain *from the outside* what someone who uses this expression or construction is up to, rather than offer anything which he could ever accept as an equivalent of this expression. Many philosophical analyses of fragments of language are such external descriptions. Moral philosophers who have held that to call something good is to express approval of it or to recommend it are saying something (rightly or wrongly) which cannot be translated into 'To be good is to be approved of' or '. . . is to be recommended'. The same goes for the comment, ' "Probably" is . . . a modal word. It affects not the content of what is said, but the way in which it is said . . .'[9] And any number of similar examples could be found. For analyses of this sort, the paradox does not even arise. There is no need at all for someone who employs a concept or a form of expression to have such an understanding from the outside of what he is up to as would make any correct description of this uninformative to him.

Some of the detailed discussions in later chapters would illustrate the general theses suggested here. Chapter 2 is concerned with a conceptual analysis of truth, in particular with what we standardly intend to record or convey when we say that something is true, and hence with the implicit logic of '. . . is true'. In itself, this would not be a long or difficult task, and much of this chapter is devoted to showing how other inquiries have tended to usurp this one's place. Chapter 3 discusses conditionals and if-sentences, giving in the end what

[9] J. R. Lucas, *The Concept of Probability*, p. 4.

I have called an external description of their function. Chapter 4 is partly concerned with analysing the concept of a dispositional property, but it also enters into the factual analysis of dispositional properties and powers and discusses ontological questions. In Chapter 5 it is argued that there are several concepts of probability, which need not only to be clarified and distinguished but also to be related to one another at least historically and in some cases logically as well. The main task that results is to distinguish sound from unsound ways of reasoning about probabilities, but here too there are ontological questions, problems of factual analysis in connection with some of the main probability concepts. Chapter 6, on logical paradoxes, is not primarily concerned with any of the varieties of analysis discussed here, but some of its results bear on the question whether such ordinary concepts as those of truth and of a class, analysed in a straightforward way, are coherent or not.

2

SIMPLE TRUTH[1]

§ 1. *The Bearers of Truth*

TRUTH is a universal, and the word 'truth' is an abstract noun. But most statements that appear to be about universals can be replaced by statements about individuals being of certain sorts, and concrete nouns and adjectives can do most of the work done by abstract nouns. We lose very little, and gain much in clarity and definiteness, if we replace the question 'What is truth?' with the question 'What is it for something to be true?'[2] However, I suspect that the high-sounding question 'What is truth?' has often been thought to be at once peculiarly important and peculiarly difficult to answer because it has been taken at least partly in a different sense, as equivalent to 'What is the truth?', that is, to 'What are (all) the true statements?' Unless it is very narrowly restricted by the context in which it is asked, this, being a demand for all the infinitely many true statements that could be formulated, is just not a sensible question to ask. By asking merely 'What is it for something to be true?' we avoid any suggestion that we are, or ought to be, dealing with this grand but impossible question.

But what is it for *what* to be true? Before we can decide what '*x* is true' means, we had better decide what sort of item this '*x*' must stand for; in other words, what are the truth-bearers, the sorts of things that we can call true?

Such uses of 'true' as 'He has now shown his true colours' and 'He is a true friend' are, from a logical point of view, peripheral, however important they may be for the history of the meaning of the word.[3] The central uses are those in which

[1] This chapter is based on my article 'Simple Truth', *Philosophical Quarterly*, 20 (1970), 321–33. A number of the papers to which I refer are included in *Truth*, ed. G. Pitcher, and will be referred to by page numbers in this collection.

[2] Cf. Austin, *Truth*, p. 18.

[3] Cf. *Oxford English Dictionary*, under *true*.

beliefs, remarks, statements, propositions, and perhaps sentences are said to be true, and descriptions are said to be true of objects. Competing theories of truth are sometimes supported by the claim that items of this or that sort are the primary bearers of truth, but it seems to me perverse to aim at any exclusive choice between the rival candidates for this role. In the end, some account will have to be given of what it is for beliefs to be true, for remarks to be true, and so on, and nothing is achieved by insisting on the greater naturalness of this or that choice of a subject for the predicate 'true'. There seem, in fact, to be systematic coincidences between the truth of certain sentences, of certain corresponding statements, and certain predicate-expressions being true of certain objects. Thus the circumstances in which the *statement* that Peter is clever is true are just those in which the *sentence* 'Peter is clever' is true, provided that in that sentence 'Peter' is being used to refer to the Peter I am talking about in *this* sentence, and that 'is clever' is being used with its standard meaning; and this *predicate-expression* 'is clever', with the same meaning, is true of Peter in just the same circumstances again. Obvious though all this is, let us try to formulate the point more generally.

To do this, we need a general term for such *objective* items as is-blue, is-clever, and so on (that is, for the sort of thing for which the expressions 'is blue', 'is clever', and so on might be said to stand), analogous to the general term 'object' as used for such objective items as Peter, this book, etc. 'Property', 'attribute', and so on as ordinarily used will not do; the property would be, say, blueness, not is-blue. I think that 'concept' as Frege used it, intended to be stripped of its psychological associations, would do the job; so would 'predicate' if it could be stripped of its linguistic associations; a clumsy but less misleading coinage would be 'is-of-a-kind'. Since to be blue is to be of a certain kind, we may say that is-blue is a certain is-of-a-kind.

Then where 'F' is being used as a predicate-expression for the is-of-a-kind (or 'concept') G, and where 'a' is being used as a name or definite description of the object b, the following will all hold together if and only if Gb:

(1) The statement-that-Gb is true.

(2) The sentence 'Fa' is true.

(3) The predicate-expression '*F*' is true of *b*. Considerations of symmetry invite us to add a fourth member to this list, which we might formulate as

(4) The subject-expression '*a*' is sub-true of *G*.

This formulation at once makes it plain that while a predicate-expression's being true of an object is in a certain respect a simpler matter than a sentence's being true, and a subject-expression's being sub-true of a concept or is-of-a-kind is also a simpler matter than a sentence's being true, a statement's being true is in this respect the simplest matter of all. What is said in (2) above is subject to both the semantic provisos about '*F*' and '*a*', what is said in (3) and what is said in (4) are each subject to one of these, but what is said in (1) is subject to no semantic provisos at all. The truth of sentences, something's being true of something, and something's being sub-true of something, are all partly semantic matters; but the truth of statements is not. This suggests that it might be easiest to start by discussing the truth of statements.

However, 'statement' is crucially ambiguous. A statement may be an act of stating something, some kind of performance, or it may be the content of such an act, what is stated. The former would bring in a great many features that are not directly relevant to truth; we should sacrifice most of the apparent advantages of talking first about the truth of statements if we then interpreted 'statement' in the former way. From now on, therefore, 'statement' will be used in the latter sense, to refer to what is stated, to a certain content.

Belief is another candidate for the role of primary truth-bearer. But 'belief' is ambiguous in much the same way as 'statement', between an act or state of believing, or a belief-disposition, on the one hand, and something that is believed on the other. If we speak of the belief that the earth is flat, we are identifying the belief with a what-is-believed, a content. But if we say that someone's beliefs arise from his early education, we are identifying these beliefs with his state of believing or his belief-dispositions: it is for these that we are offering a causal explanation. Now if a belief is true, it is so because of what is believed, and not because of any of the features that belong rather to a mental state or disposition. It is the what-is-believed, then, that is the serious candidate from the belief camp. But in all ordinary cases, something that is believed can equally be something that is stated; a belief-content is identical with some statement-content; these constitute only one sort of candidate, not two.

Several considerations, then, encourage us to start by discussing the truth of statements in the sense of what is stated. Philosophers are, however, sometimes unwilling to do this because they think that statements are ontologically suspect, or are obscure entities,[4] that they have not sufficiently precise identity-conditions, and so on, perhaps even that they are fictitious hypostatizations, whereas sentences or, still better, utterances and inscriptions, are reassuringly concrete. But these fears are largely beside the point. It is true that statements have some of the features of universals. Utterances, inscriptions, speech-episodes, people being in certain belief-states, and so on are the concrete realities, while statements are abstractions and exist not on their own but only as the contents of such speech-episodes and the like. But just as we must not push the thesis that everything that exists is particular to the extreme of denying that these particulars are, intrinsically, of certain sorts, so we must not deny that speech-episodes, believings, etc., have contents. And as a rule we can identify them well enough. Though there are some tricky cases, we know in general what it is for two people to share a belief, or for one to deny what another asserts. We normally specify such contents by using that-clauses in indirect speech—'He said that . . .', 'The belief that . . .' It may be said that this method of specification carries obscurity with it, that there is doubt as to how indirect speech is to be construed, and in particular that there are problems about referential opacity. But where what is called opacity occurs, it can be eliminated; we can replace an opaque specification of, say, a belief by a specification which contains either no expressions used referringly or only ones used transparently. And then anything formulated in indirect speech as the statement that so-and-so is just as precise as the sentence 'So-and-so' in direct speech would be. If and in so far as we understand this sentence, we can identify the content introduced by the corresponding that-clause. Such contents will remain obscure only for those who embrace a total scepticism about communication.

[4] Cf. W. V. Quine, *The Ways of Paradox*, p. 143.

The notion of propositions—in a sense practically identical with that I am giving to 'statements'—has been defended against some criticisms but subjected to others by J. F. Thomson.[5] The notion of something assertible, he says, is not *obscure*. But he objects that propositionalism provides no account (obscure or otherwise) of what it is to say something or assert something, or of what it is that is then asserted. He thinks that propositionalism would be an interesting thesis only if it defended the positive claim that 'there is some *one* proposition that snow is white' or that when a sentence is uttered there is some one proposition that has been asserted. He thinks that ambiguity, particularly unnoticed or unresolvable ambiguity, is fatal to any interesting propositionalist thesis. 'Nothing is . . . explained or clarified' when the propositionalist merely 'regards it as evident that assertions get made and gives the name "proposition" to that which is asserted.'

But this is not to the point. In order to use statements or propositions as truth-bearers it is not necessary to have a theory or to pretend to have a theory that purports to explain what it is to say something, nor to deny that a remark may be persistently ambiguous, nor indeed to do any of the things that would make 'propositionalism' interesting to Thomson. Statements or propositions are not (as, say, electrons and genes are) entities by the non-trivial postulating of whose existence we can better explain (and perhaps even predict) what goes on and is observed. The words 'statement' and 'proposition' are just terms that enable us to speak generally about what is said, what is believed, what is assertible or believable, and so on. Of course an utterance may be ambiguous and the speaker may not have intended one or other of the possible interpretations; of course there may be ambiguities which we have still not noticed in the sentence which someone used and which we borrow when we report 'He said that . . .'. But *if* there is an ambiguity, there is a corresponding fuzziness about the issue whether what he said is true.

None of the ordinary objections, then, tell against the choice of statements or propositions as the primary truth-bearers. There are, however, other and perhaps better reasons for

[5] J. F. Thomson, 'Truth-bearers and the Trouble about Propositions', *Journal of Philosophy*, 66 (1969), 737–47, esp. 740, 742, 745.

concentrating on the truth of sentences instead; I shall consider these later (in § 3).

If we do take statements to be the primary bearers of truth, there seems to be a very simple answer to the question, what is it for them to be true: for a statement to be true is for things to be as they are stated to be. Aristotle's dictum is well known: '. . . to say of what is that it is, or of what is not that it is not, is true.'[6] This is no doubt too narrow, since there are many cases where what is said is neither that something is nor that something is not. But we can retain the notion but generalize it, saying 'For any *p*, to say that *p*, where *p*, is true'. The truth-condition for anything introduced as the statement, belief, and so on, that *p* is simply *p*. And to say that a certain statement *S* is true is to say that, for whatever *p* we can identify *S* as the statement that *p*, *p*.

In the end, I think, we shall be able to accept this. But before we do so we must deal with a number of objections and rival views.

§ 2. *Pragmatist and Coherence Theories of Truth*

Why have philosophers not been content with such a very simple account of truth as this? Why do we encounter a number of rival theories of truth?

One reason is that philosophers have wanted a theory of truth to do more than this; not just to say in a broad way what it is for a statement to be true and what we are saying when we call a statement (or sentence or belief or utterance and so on) true, but to provide a criterion of truth, a set of rules or a standard procedure by the application of which we can decide, in each particular case, whether a statement (or sentence etc.) is true or not. But another reason is that philosophers have often doubted whether simple truth was within our power, whether we could succeed in saying how things are, in saying that *p* where *p*. Some of the rival theories of truth are essentially sceptical. Thus[7] the logical positivist variant of the coherence theory of truth had its root in a belief that it is metaphysical to talk of comparing statements with facts: all we can do is to

[6] Aristotle, *Met.* 1011b 26ff.

[7] As A. J. Ayer explains in 'Truth', *The Concept of a Person*, p. 179.

compare statements with statements, and coherence among statements is then offered as an account of truth because it is thought that this is the most we can achieve. Simple truth, saying that things are as they are, is beyond us. A pragmatist theory of truth may be adopted for similarly sceptical reasons. We are not justified in claiming to describe things as they are, but we may succeed in making statements that are useful for certain purposes. It is highly plausible to speak in this way about scientific theories and hypotheses, which enable us to make predictions that are fulfilled and to construct machines that work but which are not merely fallible in principle but also very likely to be falsified eventually and to be either modified or replaced by new theories. We may be tempted to assign to all of what we call our knowledge the same status that we give to scientific theories; if so, we may well conclude again that simple truth is beyond us, and that the most we can reasonably claim for any of our statements is that they are useful.

The reasons for which William James adopted the pragmatist theory of truth were different again. He criticized the 'copy' version of the correspondence theory, and he did not sufficiently distinguish *what* we are saying (when we say that something is true) from our reasons for saying it. But above all he thought that what the pragmatist theory calls truth is more interesting and important than what is so called by the 'intellectualist' theory.[8]

Thus understood, such rival theories of truth are not disputing the simple or classical analysis of the ordinary concept of the truth of statements, they are not denying that the simple account gives the ordinary meaning of the word 'true' as applied to a statement. They may well concede that this analysis elucidates what we commonly intend to assert when we say that some statement is true. But they suggest that this meaning is out of place, that we are either not entitled to assert what we commonly intend to assert, or wrong in thinking that this is much worth asserting. They therefore *propose* that the word 'true' should change its connotation, and perhaps argue that it must do so if it is to keep its denotation unchanged. The word 'true' should be given a different job that is more practicable or more important. Alternatively, or perhaps additionally,

[8] W. James, *Pragmatism*, Lecture VI.

such rival theories may attempt to supply not merely an analysis of the meaning of 'true', but a criterion of truth, a general procedure for finding out whether a statement is true or not.

A purely analytical argument is powerless against rival theories understood in this way: to protest or even to prove that some simple classical account correctly analyses the ordinary concept of truth is, in this context, to miss the point. If we want to challenge such rival theories we must meet them on their own ground.

And so we can. Let us take first the demand for a general criterion of truth. It is clear that the suggestions most frequently put forward are inadequate as criteria of *truth in the ordinary sense*. Coherence, for example. It is sometimes thought that this is a criterion of truth at least in formal systems of mathematics and logic. We do sometimes say that a proposition or theorem is *true within a certain system*, say Euclidean geometry, meaning just that it is a theorem of that system, that it is provable in the sense that it can be derived from the axioms of the system in accordance with the transformation rules of the system. This, however, is a technical way of speaking, and a misleading one.[9] Also, this concept of truth within a system has a very different logic from the ordinary concept of truth. Ordinary truth obeys the law of excluded middle; as soon as we have adequately specified a statement and its negation, one or other of them must be true. But it may well be that of a well-formed formula of a certain system, and its negation, neither is a theorem of the system, neither can be proved within it. In any case, this concept of truth within a system has no application where there are neither axioms nor transformation rules laid down; if 'coherence' is to be a general criterion of truth it must be interpreted in some much wider sense, either negatively as the mere absence of logical inconsistencies between the statements of a set, or positively as the presence of some relations of support, perhaps mutual support, between some or all of the statements of a set or the fact that the statements of a set somehow constitute an aesthetically complete and rounded picture. It is going to be difficult to make any such positive notion of coherence precise enough to provide a criterion. Even if this

[9] Cf. my 'Proof', *Aristotelian Society Supplementary Volume*, 40 (1966), 23–38, esp. 23–4.

were done, we should still have to say about such positive coherence what it is abundantly clear that we must say about negative coherence (the absence of inconsistencies), that it is not a sufficient condition of truth. A set of statements or beliefs can be as coherent as you like, and yet the world may fail to be as they describe it. We must concede that we are never able to reject a statement or set of statements as being (at least partly) false until we encounter some inconsistency either within it or between it and something else that we observe or believe or are inclined to accept. A set of beliefs which was at some time completely coherent in this sense of being neither internally nor externally in conflict would be accepted at that time as being true. But its coherence does not establish its truth; it only gives it an as yet unchallenged claim to the title of truth. Even when coherence functions in this way as a temporary criterion of truth, it does not work on its own. Within any body of beliefs that are held by someone at some time there are some which he accepts partly because things just seem to be as he believes, not merely because these beliefs fit in positively or negatively with others. While coherence is neither a sufficient nor an independent test of truth, it is worth noting that although negative coherence is at least a necessary condition for truth, positive coherence is not even this; an isolated statement or belief that is in no way supported or aesthetically complemented by others may well be true in the ordinary sense.

Equally, pragmatism cannot provide either a necessary or a sufficient criterion for truth in the ordinary sense. Even if we interpret 'usefulness' as shorthand for success in relation to the whole practical and scientific procedure by which hypotheses are tested, it is clear that this never constitutes a criterion of truth. An hypothesis that has been thoroughly tested and has so far survived and proved satisfactory is confirmed or corroborated, and we may for the time being treat it as true, call it true, perhaps even hope that it is true; but we know perfectly well that it may later turn out not to be so. Conversely, an hypothesis that has not stood up well to testing, that has apparently been falsified, may be resurrected and re-established with the help of some change elsewhere in our theories. What is 'useful' may not be true, and what is not, for the moment, 'useful' may be true after all. Again what we have is a criterion not of truth but of something's being entitled to be called true, temporarily and provisionally.

Similarly we can rebut the thesis that simple truth is beyond our power. As Ayer says,[10] observation does enable us to break out of the circle of statements. Seriously to deny this would commit us to an extreme scepticism in the light of which the making of empirical statements would be a singularly pointless activity. Again, the pragmatist version of scepticism is made plausible by the conflation of two theses. Scientific theories and hypotheses are both fallible in principle and fairly likely to need revision as time goes on. Everyday judgements about matters that are open to fairly direct observation are also fallible in principle. Though ordinary language philosophers have been inclined to deny this, I think it is sensible to say that both particular statements about, say, the number of chairs in a room, and such very general theses as are incorporated in the commonsense view of the world as consisting at least partly of material things, spatially ordered and persisting but also changing through time, are in principle fallible and have approximately the same logical status as well-confirmed hypotheses. But in conceding this we do not need to concede that these particular judgements and general structural theses are actually likely to need revision as time goes on in the way that scientific theories are, which as it happens are intermediate in generality between these two classes of commonsense claims. Both kinds of commonsense claims are fallible; they too are confirmed rather than conclusively verified; but none the less they are very likely to be right, and it is not at all likely that we shall need to scrap them or modify them. So although it is rather rash to claim simple truth even for such well-confirmed scientific theories as the gene theory of inheritance or special relativity, it is not at all rash to claim simple truth for a careful report, checked by a second observer, about the number of chairs in a room, or for the anti-Berkeleian thesis that material things exist independently of minds. In some very obvious areas simple truth is within our power.

This conclusion can be supported by an *ad hominem* argument. Anyone who denies this, but says that some form of 'usefulness' is within our power, can be convicted of implicitly claiming

[10] Ayer, op. cit., pp. 180, 186.

simple truth for some of his own statements, e.g. for the statements that a certain theory *is* useful, that certain predictions have been fulfilled, that this or that hypothesis has been corroborated. But the simple truth of such statements as these involves and presupposes the simple truth of everyday statements of kinds which the sceptic was assimilating to the propositions of scientific theory for which it seemed rash to claim simple truth.

Reluctance to accept this conclusion is sometimes due to a confusion between truth and certainty. It may be thought that if we say that a statement is *true*, and not merely useful or well confirmed, we are saying that it has been proved or established beyond reasonable doubt. If this were so, then the claim that simple truth is within our power would conflict with the fallibility that I have already conceded. But to claim truth in the ordinary sense is not to claim certainty. Again we can point to a difference between the logical rules for the two concepts. A statement and its denial may both fail to be certain, but they cannot both fail to be true. To assert that some statement *S* is true is no more rash than to assert *S* itself, but to assert that *S* is certain is to make a further claim. Admittedly one of the uses of the statement that some statement *S* is true is to re-assert *S* or emphatically to adhere to it in the face of some doubt or challenge; '*S* is true' can be an expression of certainty. But it does not follow that '*S* is true' means the same as '*S* is certain', and in fact it clearly does not. Simple truth need not be blown up into certainty, any more than it needs to be reduced to usefulness or confirmation; but it is sometimes thought that the second of these is the only alternative to the first.

We need not waste much time over the suggestion that the simple concept of truth is uninteresting or unimportant, and that the word 'true' should therefore be taken over as a name for some more exciting feature. No doubt many sorts of issue may call for attention, but whether things are as they are said to be is frequently of some significance; we need a word to do precisely the job that 'true' in its central ordinary sense already does.

§ 3. *Correspondence and Semantic Theories of Truth*

Traditionally it has been assumed that if what I have been grouping together as the various rival theories are out, then a correspondence theory of truth is in. But several forms of correspondence theory have been proposed and criticized in turn, and the currently dominant view seems to be that while we want the concept of objective truth, or a realist theory of truth, we do not want a correspondence theory.[11]

There are two problems for any correspondence theory: to identify the items which are supposed to correspond where there is truth but to fail to correspond where there is falsehood, and to say in what the correspondence or non-correspondence consists. Characteristic failings of such theories are that one of the two items becomes a mere shadow of the other and cannot be separately identified, and that the relation called correspondence remains hopelessly obscure. Another well-known objection —whether it is a failing or not remains to be seen—is that any correspondence theory improperly mixes up questions of meaning with the question of truth.

Russell, between 1906 and 1912, put forward one form of correspondence theory.[12] Though he himself rejected it later and no one now, I think, takes it very seriously, it is worth considering for the light it throws on the general problem. Insisting that an adequate theory of truth must leave room for an account of falsehood also, and arguing that where a belief is false there is neither a state of affairs nor a proposition-entity that can be taken as the second term of a relation of believing, he took as one of the items that may or may not correspond a concrete state of believing construed as a many-termed relation between a believer and whatever might have been taken as the objective components of the fact that would have been there if the belief had been true. Taking his own example, and treating it as if the story Shakespeare told were historically accurate and not a mere fiction—since allowing for the fiction would introduce unwanted complications—we find that Othello's believing that Desdemona loves Cassio is treated as a rather complex relation between Othello, Desdemona, Cassio, and the (universal) relation *loving*. If this belief had been true there would have been also a rather simpler relation between

[11] Cf. M. Dummett in *Truth*, ed. G. Pitcher, p. 106: 'Baffled by the attempt to describe in general the relation between language and reality, we have nowadays abandoned the correspondence theory, . . . but we remain realists *au fond*; we retain in our thinking a fundamentally realist conception of truth.'

[12] B. Russell, *Problems of Philosophy*, ch. 12.

Desdemona, Cassio, and *loving*, namely the situation describable by 'Desdemona loves Cassio'. The belief's being true would then have consisted in the partial formal similarity between the belief-relation which holds for the ordered set (Othello, Desdemona, *loving*, Cassio) and the relational situation, Desdemona-loving-Cassio. Its falsehood consists in there not being any relational situation to which this belief-relation has the partial formal similarity in question. The fundamental difficulty for any such account is that it is not plausible to take this state of believing as a relation between Othello, Desdemona, etc., though no doubt the previous or present holding of causal and/or cognitive relations between Othello and Desdemona etc. is a prerequisite for there being such a state. Othello's believing that Desdemona loves Cassio is, as it exists at the time, something intrinsic to Othello, it does not have as components the other individuals Desdemona and Cassio, and it is quite obscure how it could have as a component the relational universal *loving* which (since Othello's belief is false) is not instantiated here at all. But even if this difficulty did not arise, even if we could construe Othello's believing as some such complex relation, we still could not say how this relation would correspond to Desdemona's loving Cassio, if that were realized. How is the direction of the latter relation—the fact that it would be Desdemona's loving Cassio, not Cassio's loving Desdemona —mirrored in the more complex belief-relation? But if we abandon the complex relation analysis of believing, it becomes still harder to explain any correspondence. At least the supposed belief-relation between Othello, Desdemona, *loving*, and Cassio had the merit of sharing components with the state of affairs (had it occurred) Desdemona-loving-Cassio. There is the beginning of a partial formal similarity between them. But if we correct our account of Othello's believing so that it is no longer a relation between these components and Othello, we cannot even begin to say how Othello's inner state can be mapped onto the external state of affairs. There is no need to be intolerant or dogmatic about this. It may well be that if Desdemona had loved Cassio, there would have been some sort of systematic correspondence between this state of affairs and whatever it was that was going on in Othello when he believed this. There may be a correct correspondence theory to be discovered about what is in fact involved in a state of believing as compared with the state of affairs that makes the belief true. All I am saying is that *this* correspondence, if any, is not at present known; it cannot be what we are asserting if we assert that Othello's belief is true or denying when we say that his belief is false. The sort of analysis that would uncover *this* correspondence, if it exists, is an empirical inquiry into the details of what goes on, and would be quite different from the more obviously philosophical analysis of the concept of truth.

What led us into this trap was an unwise choice of a primary truth-bearer. It was in looking for a relation between the *state of believing* and what would make it true, rather than between the latter and *what is believed*, that we started talking about forms of correspondence that are, at present, quite unknown.

Of far more permanent importance is Tarski's 'semantic' conception of truth,[13] which can be considered as a kind of correspondence theory or as a development from a correspondence theory. Tarski chooses sentences as truth-bearers, and his aim is to 'define "true sentence" ' in the sense of clearly marking off true sentences from the rest, that is, to provide a criterion for the truth of sentences rather than an account of what 'true' *means* even as applied to sentences. This starts from such examples as that the sentence 'Snow is white' is true if, and only if, snow is white. Of course it is not the sentence 'Snow is white' absolutely that is true on these conditions, but 'Snow is white' as an English sentence. More generally, any formula such as ' "Snow is white" is true if, and only if, snow is white' will be correct only provided that the quoted sentence 'Snow is white' is being used in the same way in which the same sequence of words is used after 'if, and only if'. The point would be even clearer if the quoted sentence included proper names and/or egocentric or indexical terms. The sentence 'I was at home yesterday', as used by Smith on Tuesday, is true if, and only if, Smith was at Smith's home on Monday—that is, the home of the same Smith and on the right Monday. And similar provisos would be needed for 'Smith was at home yesterday'. What is true is never a sentence absolutely, but a sentence as used with

[13] A. Tarski, 'The Concept of Truth in Formalized Languages', *Logic, Semantics, Metamathematics*, pp. 152–278; 'The Semantic Conception of Truth', *Readings in Philosophical Analysis*, ed. Feigl and Sellars, pp. 52–84.

certain meanings and perhaps certain references for its words and phrases. But with suitable provisos of this kind, it would be tempting to generalize from the above example and say 'For all *p*, "*p*" is a true sentence if and only if *p*'.

Tarski has two reasons for not following this suggestion. One is that if this rule is applied without further qualification to grammatical sentences in a natural language such as English, it produces contradictions such as the paradox of the liar. If we use '*c*' as an abbreviation of some definite description of the sentence '*c* is not a true sentence', the proposed rule would yield the contradiction '*c* is a true sentence if and only if *c* is not a true sentence'. The conclusion Tarski draws is that a consistent language cannot contain its own semantics, that if a language does include names for its own expressions, and 'semantic' terms such as 'true', and one can formulate within it adequate rules for the use of such terms as 'true', and if when using it one can use the ordinary logical rules, such contradictions will be unavoidable within it. He further concludes, therefore, that we must distinguish sharply between the object language for the truth of whose sentences we are trying to provide rules and the metalanguage in which those rules are formulated. That is, even in the elementary example, where the word-sequence 'Snow is white' occurs first, it must be taken to be a sentence of the object language, but its second occurrence is as a clause in a sentence in the metalanguage.

I shall be discussing such logical paradoxes and this way, among others, of handling them in Chapter 6. But for the present I concede that if what we want is a set of rules for determining the truth of sentences that shall be mechanically applicable without restriction, then the liar paradox does show that such a set of rules cannot be stated within the language to which it is intended to apply (if it also has the above-mentioned features). But we could still say 'For all *p* except some very queer ones, "*p*" is a true sentence if and only if *p*'.

Tarski's second reason for not saying this is a difficulty about reading the symbol ' "*p*" ' in the middle of this formula. He first suggests that this must be taken as a name of the letter '*p*', so that ' . . . "*p*" is a true sentence . . .' will be nonsense. But it is surely perverse to let ourselves be bullied by symbols. We can make our symbols serve our purposes, and the suggested formula can be read so as to be a generalization each of whose instances has some name of a sentence before the words 'is a true sentence if and only if' and the same sentence itself after them. Tarski recognizes this possibility, calling such expressions as ' "*p*" ', read in this way, quotation-functions. But he still complains that their sense is not sufficiently clear—a charge which I do not understand—and that they too expose us to the risk of paradoxes. As far as I can see, then, paradoxical constructions constitute the only substantial difficulty for the proposed generalization, and they are sentences of a very special sort.

With some reservations about paradoxical sentences, then, and subject to the above-mentioned provisos about meaning and reference, I think this generalization would give a correct account of the circumstances in which sentences are true. It does not yet say explicitly what 'true' means when applied to sentences, but an account of this could fairly easily be developed from it. Where '*S*' is any name (or definite description) of a declarative sentence of the language we are presently using, to say that *S* is true is to say that for whatever word-sequence '*p*' *S* consists of, *p*. That is, to say that *S* is true is to say that if *S* is 'Snow is white' then snow is white, and that if *S* is 'Chimpanzees can talk' then chimpanzees can talk, and so on. This will be correct subject to the usual provisos about the two uses of the word-sequence here represented by '*p*'.

Even if correct, however, these formulations are not very interesting. Each formula of the form ' "*p*" is a true sentence if and only if *p*' will inform someone about the truth-conditions for '*p*' only if he knows already how to use '*p*' in the same way, that is, if he already knows the truth-conditions for this '*p*'. It tells him no more than he knows already. But it is possible to construct a systematic account of the truth-conditions for the sentences of a particular language which is not trivial in this way. And this is the line of thought Tarski follows.

If *S* were a truth-functional compound of other sentences, e.g. if it had the form '*A* & *B*' or '*A* ⊃ *B*', then we could give its truth-conditions in terms of the truth and falsehood of its components. So if a language *L* consisted only of a set of elementary sentences and truth-functional compounds we could give a 'definition of truth for this language', i.e. a set of rules for determining the truth or non-truth of all sentences of that language, by giving 'Snow is white' type rules for the truth of each elementary sentence together with the truth-functional rules. That is, we could 'define' '*S* is true in *L*' somewhat as follows:

Either (*S* is 'Snow is white' and snow is white) or . . . or (*S* is of the form *A* & *B* and *A* is true in *L* and *B* is true in *L*) or (*S* is of the form $A \supset B$ and either *A* is not true in *L* or *B* is true in *L*) or . . .

This would be a directly recursive definition of the truth of sentences. But, as Tarski points out, this procedure is not available where compound sentences are constructed from simpler sentential functions (containing variables) and not only from simpler sentences. He therefore—in order to provide a 'definition of truth' for formalized languages that contain predicates, variables, and quantifiers—approaches truth indirectly by way of what he calls *satisfaction*. For example, the ordered pair of numbers (7, 4) satisfies the function '*x* is greater than *y*'. The satisfaction of more complex sentential functions by sequences of objects is first defined in terms of the satisfaction of simpler functions—that is, a recursive definition of satisfaction is built up, doing for the satisfaction of sentential functions what we could have done for the truth of sentences in a language whose complexity was merely truth-functional. Then sentences are taken as sentential functions with no free variables, so that a sentence is satisfied either by all sequences of objects or by none. And finally truth is defined for sentences by saying that a sentence is true if it is satisfied by all (sequences of) objects, and false otherwise.

The technical details of Tarski's procedure, even for the formalized languages with which alone he is concerned, are extremely complex; but the questions of philosophical interest are about what it achieves, what its point and value are. Those who are competent to judge agree that he achieves what he sets out to do, to supply rules which entail, for the languages in question, all the 'equivalences of the form (*T*)', that is, all the statements of the form

X is true if, and only if, *p*

in which '*X*' is replaced by a name or definite description of a sentence of the object language and '*p*' by that sentence itself —assuming that it is also a sentence of the metalanguage. In other words, his definition yields, by its unpacking, the desired answers which were given immediately by our not very interesting formulations above, while the unidirectional recursive procedure ensures the finite eliminability of 'semantic' terms, and so provides 'a kind of guarantee' against the occurrence of self-referential paradoxes.

But apart from the avoidance of the paradoxes, what is the merit of Tarski's very complex procedure as opposed to our rather trivial one? Clearly what extra a Tarskian truth-definition supplies is an account of the structural meaning of the language in question, or, more accurately, an account of the contribution which the internal structure of complex sentences of the language makes to their meaning. Equally clearly, this is additional information of great interest. And yet there is room for several queries.

First, is a 'truth-definition' the best form in which to present this information? Did not Tarski already possess all this information in a more compendious and more easily handled form before he set out on his task of construction, namely in a collection of semantic rules about the various ways of compounding sentential functions, using quantifiers, and so on? Is not the project of constructing *a* truth-definition for a language an unduly grandiose one as compared with the task of explaining, one by one, the contribution that each of the various forms of composition allowed in the language makes to the meaning (by way of truth-conditions, satisfaction-conditions, or whatever) of the compounds it helps to produce?

This is only a pragmatic point, and I am content to leave it as a query. More substantial is a criticism which Strawson has put forward on many occasions, that the semantic theory, and indeed every form of correspondence theory which identifies truth with some relation between language and the world, mixes up questions of meaning with questions of truth and is thereby committed to an inadequate treatment of meaning.[14] But is this so? We must concede that some branches of linguistic meaning do not lend themselves to exposition by

[14] P. F. Strawson, 'Truth', *Philosophy and Analysis*, ed. Margaret Macdonald (reprinted from *Analysis*, 9 (1949)), and his contributions to *Truth*, ed. G. Pitcher.

truth-conditions—imperatives, for example, and questions, perhaps evaluative statements, emotive language, and (as I shall argue in Chapter 3) non-material conditionals and perhaps dispositional statements (see Chapter 4) and some kinds of probability statements (Chapter 5). Yet for some of these some analogue of a truth-conditional approach may be in order: we might explain imperatives in terms of their fulfilment-conditions, and at least the greater part of the meaning of a question is displayed by correlating it with a (finite or infinite) set of allowed answers.[15] But where and in so far as sentences admit of truth and falsehood there is necessarily a very close connection between their meaning and their truth-conditions. The primary task of any declarative sentence in use is to convey some claim about how things are. This will be its basic meaning, whatever further illocutionary or perlocutionary functions it may have as well. So we are inclined to argue as follows.

The basic meaning of a declarative sentence both fixes and is fixed by how things are said to be when it is used assertively. A sentence as used will be true if, and only if, things are as they are then said to be. So to fix the truth-conditions of a sentence as used will be to fix its meaning. But nothing in this mixes up truth and meaning. Considerations of meaning (which may include conventional meaning-rules, or the speaker's intentions, or both) link the sentence as used with how things are said to be. Truth (or falsity) lies in the agreement (or disagreement) between how things are said to be and how they are, as was suggested at the end of § 1. These two distinct and heterogeneous links form a chain which ties meaning to conditions for truth. Strawson suggests that there is a special metalinguistic use of 'is true if and only if' which makes this phrase synonymous with 'means that', and this (in each of Tarski's equivalences of form (*T*)) is mistakenly taken as a model for the general use of 'is true'.[16] But this is not a special use, and there is no mistake or confusion. A natural extension of its basic general use allows 'is true' to be predicated of a sentence, with the meaning 'is used

[15] Compare 'Is he in or out?' with the set of allowed answers ('He is in': 'He is out'); and 'When was oxygen discovered' with 'Oxygen was discovered in . . . (date)'; and so on. Stated alternative and yes-no questions correspond to finite sets of allowed answers, *W*-questions to open sets of allowed answers with category restrictions only on how the gaps are to be filled.

[16] Strawson, 'Truth', *Philosophy and Analysis*, pp. 263–6.

to make a true statement'. It is then a consequence of the basic meaning of 'true' that *S*-will-be-true-if-and-only-if-*p*—no matter how the world varies—if and only if *S*-means-that-*p*. Thus 'is true if and only if' is not synonymous with 'means that', but pairs of sentences of these forms will regularly be materially equivalent as a result of the central sense of 'true'.

This argument will not *quite* do, for two reasons. Although the basic meaning of a sentence as used fixes how things are then said to be, the converse does not universally hold. Two sentences of different internal structure may issue in the same claim about how things are, but because they have reached the same goal by different routes, it is natural to say that their meanings are different. Also, what a sentence as used presupposes and what it asserts over and above that presupposition play different roles in its meaning, but if we attend only to truth-conditions, this difference may be obscured. To fix the meaning, we may need to specify not only the truth-conditions but also the falsity conditions and the-question-doesn't-arise conditions. Two sentences might have the same truth-conditions, but if one was false in circumstances in which the question of the other's truth or falsity did not arise, they could not have the same meaning. However, neither of these points is of great importance for the present issue. The first would tell only against a direct truth-conditional account of meaning, not against a Tarskian approach in which structural meaning is revealed by the step-by-step procedure by which satisfaction, and only ultimately truth, are tested. And the second distinction, while not relevant to the sorts of language with which Tarski was concerned, could easily be incorporated in a procedure of the same general kind. The important part of our argument still stands: we can rebut the Strawsonian accusation of mixing up truth with meaning by distinguishing the two links in the chain which connects the meaning of a sentence with its truth-conditions.

Yet to draw this distinction is to go against both the letter and the spirit of Tarski's procedure. He wants sentences, not statements or propositions, as truth-bearers, and he does not leave room for any distinct intermediate item or stage in either the definition of truth or the resultant truth-checking procedure, how-things-are-said-to-be. His sentences (which are of course 'eternal' sentences, with no egocentric or indexical terms) are themselves directly to be tested for truth. Our two links are fused in Tarski's approach. And this does seem to be a fault.

The same point may be made in another way. What are we saying when we ascribe truth to a sentence in use? Not, surely, that a Tarskian truth-predicate holds of it, for if we ordinarily ascribe truth to sentences (in use) it is to sentences of natural languages, and Tarski himself abandoned as hopeless the task of defining truth for a natural language. But what we are saying is something neither complex nor obscure, but just that however things are, by the use of this sentence, said to be, so they are. But this simple, everyday, sense of 'true' can be displayed only by using the notion of how things are said to be, for which the Tarskian approach leaves no room. It would be a merit of a theory of meaning and truth if it did leave room for this.

A defender of Tarski might make two replies to this. One is that the everyday sense of 'true' for which I am trying to leave room is itself incoherent and has to be abandoned. I think that this is not so, but since the case against the everyday sense rests mainly on the logical paradoxes, this issue must be postponed until Chapter 6. The other possible reply (not, as far as I know, actually favoured by Tarski or his followers) would be to offer an analysis of the ordinary sense of 'true' as applied to declarative sentences along the lines I suggested above: to say that *S* is true is to say that for whatever word-sequence '*p*' *S* consists of, *p*. This attempts to give a general simple account of the meaning of '*S* is true' in terms only of *words*, avoiding the elusive what-is-said or how-things-are-said-to-be. But this account is not yet general enough. 'True' in the ordinary sense can be applied by a speaker to sentences in a language other than the one he is using, so that we should have to insert 'or a translation of *S* into the language we are now speaking' between '*S*' and 'consists of' in the above analysis. And then if we want our account to cover sentences containing egocentric or indexical terms, we should have to incorporate rules providing for the appropriate transformations between the original sentence *S* and the final word-sequence '*p*'. We could in this way construct an account of the truth of sentences in terms only of words, translations, and indexical transformations which would coincide with the ordinary use of 'true' as applied to sentences and would exclude the analysis of structural meaning which a Tarskian 'truth-definition' provides. But in getting this account correct, we should have lost its initial simplicity, and with the simplicity much of the plausibility of the claim that this is an analysis of the, or even an, ordinary sense of 'true'. The ordinary, simple, notion of truth involves a comparison between how things are and how they are said to be, not between word-sequences even subject to appropriate controls on sense and reference. What may confuse us is that it is only by using words, with due attention to their sense and reference, that we can specify either how things are or how they are said to be; but the words are the instruments, not the subject of our inquiry.

Let us consider again whether it is or is not truth in the ordinary sense that is introduced at some stage or another in Tarski's procedure. We must concede the correctness of each sentence of the form ' "*p*" is true if, and only if, *p*' where 'true' has its ordinary sense and the sentence symbolized by '*p*' is used with the same sense and reference in both places. But the correctness of a sentence of the form ' "*p*" is ϕ if, and only if, *p*' does not ensure that 'ϕ' means true in the ordinary sense. Even if there were a term 'ϕ' such that every sentence of this form was correct, there would be no guarantee that this 'ϕ' meant true: 'ϕ' might still be some other predicate which happened to be coextensive with 'true'. Such an indefinitely extended accidental coincidence would be highly improbable, but not impossible. Consequently, a stipulative definition for the use of 'ϕ' in the form of a rule that '*p*' is (to be) ϕ if, and only if, *p* would still not ensure that 'ϕ' meant true. It follows that we cannot analyse or define or introduce *true-in-the-ordinary-sense* by saying that '*p*' is true if, and only if, *p*. There would always be the risk of introducing not 'true' but some coextensive term, of a kind of naturalistic fallacy. It was to avoid this that I suggested that to constitute a definition of 'true', our rule should be recast in the form 'To say that *S* is true is to say that for whatever word-sequence "*p*" *S* consists of, *p*'.

It might be suggested, however, that though the rule ' "*p*" is ϕ if, and only if, *p*' would not in itself ensure that 'ϕ' meant true, this would be ensured if the rule that prescribed each case of this form were not stated directly but enshrined in a recursive definition, from which each case was derived indirectly. But there is no magic in recursiveness as such to tie a 'ϕ' it generates in this way to the ordinary sense of 'true'. There is no reason why, if some other term happened to be coextensive with 'true', it should not be open to a rival recursive definition of its own. But perhaps the *particular* recursive definition suggested by Tarski does ensure this, because there are systematic links between its 'truth' and satisfaction, and between complex and simple cases of satisfaction, which tie the defined 'truth' back ultimately to a predicate's belonging to a subject? In a sense, what Tarski's procedure introduces is a *part* of the ordinary notion of truth. Only a part, because it is essential to this procedure that truth is introduced piecemeal, for this or that 'language', whereas it is equally essential to the ordinary notion of truth that it is unrestricted, that sentences in all languages may be true in the same sense. Tarski's procedure in effect spells out how things are said to be in a standard use of a certain class of sentences, and identifies the truth of any such sentence with things being so.

Another way of avoiding what I have called the naturalistic fallacy about truth, the risk of identifying 'true' with some merely coextensive term, is to use modal terms. If we say that 'ϕ' is the predicate such that it is *necessary* that 'p' is ϕ if, and only if, p, or, more explicitly, such that if S consists of the word-sequence 'p', it is necessary that S is ϕ if, and only if, p. With a sufficiently strong sense of 'necessary', we could thus rule out even infinite coincidences, and ensure that 'ϕ' meant true, and was not merely coextensive with 'true'. But we still have to provide that the two occurrences of 'p' should have the same sense and the same *reference*, and, if they include egocentric terms, be used from the same point of view. But given all these provisos, it does look as if we can thus get something very close to the ordinary sense of 'true'. Truth is that feature which a sentence-in-use *necessarily* has if, and only if, that same sentence in that same use. I do not think that this is *quite* the ordinary sense of 'truth', just because I believe that truth in the ordinary sense applies primarily to what is said, and not even to a sentence-in-use. But those to whom statements are anathema are welcome to use this as an approximation to what they cannot accept.

It is a curious, though not objectionable, feature of Tarski's procedure that although it aims at an account of the truth of *sentences* it works in terms of the true-of relation which holds between *predicate-expressions* and objects. 'Satisfies' is merely the converse of 'is true of'. When truth is finally defined as a special case of satisfaction, sentences are in effect being treated as a sub-class of predicate-expressions, which are true of either all objects or none. What is being developed is not so much a correspondence theory of truth as a predication theory of truth.

A significant parallel can be drawn between the early Russell way of developing a correspondence theory and Tarski's semantic theory of truth. The former attempted to locate truth in a correspondence between a state of believing and an objective situation, and our conclusion was that there might be some such correspondence, but it is at present quite unknown; the study of it would belong to factual rather than conceptual analysis, and we do not need to go into this in order to discover what we are ordinarily *saying* when we say that a belief is true. The latter locates truth in a satisfaction-relation between sentences considered as a sub-class of predicate-expressions and objects in general; here we must concede not merely that there might be but that there is such a relation, that (at least for sentences in formalized languages) it is not unknown but has been discovered by analysis of the structure of the languages in question; but again that we need not go into this in order to discover what we are ordinarily *saying* when we say that a statement, or even a sentence, is true. The Tarskian theory still belongs to factual rather than conceptual analysis, but this may be partly concealed because the facts in question are facts about languages and, in the cases most amenable to Tarskian treatment, deliberately constructed languages. But that is why Tarski's theory has plenty of meat to it, whereas a correct conceptual analysis of truth has very little.

Tarski thought that his method could not be applied directly to the problem of truth in natural languages just because they have no exactly specified structure. The most that could be achieved was an approximate solution, 'replacing a natural language (or a portion of it in which we are interested) by one whose structure is exactly specified, and which diverges from the given language "as little as possible" '.[17] More optimistically, Donald Davidson has taken up the project of constructing a truth-definition in Tarski's style for a natural language.[18] Davidson claims to be defending a version of the correspondence theory: 'I think truth can be explained by appeal to a relation between language and the world, and that analysis of that relation yields insight into how, by uttering sentences, we sometimes manage to say what is true.'[19] He tries to show the need for a theory of this type of criticizing simpler formulations like those which I have endorsed. I shall consider these criticisms later (in § 5); just now I shall discuss the outlines of his positive proposals.

Davidson argues that the difficulty of attaching truth to sentences (which has often been a motive for choosing propositions or statements as truth-bearers) can be met by treating truth not as a property of sentences but as a relation between sentences, speakers, and dates.[20] The truth-predicate to be defined will be a three-place one, applying to ordered triples each consisting of sentence, speaker, date. The theory can then cope with sentences that contain indexical terms. Just as Tarski takes it as a condition of adequacy of his truth-definition that it should entail, for the language in question, all equivalences of the form (*T*)

X is true, if, and only if, *p*,

so Davidson says that an acceptable theory (for English) must entail a true sentence of the form

Sentence *s* is true (as English) for speaker *u* at time *t* if and only if *p*

no matter what sentence of English is described by the canonical expression that replaces '*s*'. But whereas what replaces '*p*' in Tarski's formula can be the same sentence of which '*X*' is a name (if the metalanguage contains the object language) or a translation of it into the metalanguage, what replaces *p* in Davidson's schema must be more complicated, it must take account of the rules that govern the use of indexical terms.

[17] 'The Semantic Conception of Truth', *Readings in Philosophical Analysis*, p. 58.

[18] D. Davidson, 'True to the Facts', *Journal of Philosophy*, 66 (1969), 748–64; also 'Truth and Meaning', *Synthese*, 17 (1967), 304–21.

[19] 'True to the Facts', p. 748.

[20] Op. cit., p. 754.

In other respects, however, Davidson follows Tarski. In particular, he proposes first to give a recursive definition of the satisfaction of open sentences (Tarski's sentential functions) by functions which assign entities to variables (an open sentence is satisfied by a function if it is true of the ordered set of entities assigned to its variables by that function), satisfaction being relativized to speakers and dates. Then a closed sentence is true if it is satisfied by all functions; truth will be consequentially relativized in the same way.

Davidson admits that '*s* is satisfied by all functions' will not mean exactly what we thought '*s* corresponds to the facts' meant; his point is rather that the two phrases 'both intend to express a relation between language and the world, and both are equivalent to "*s* is true" when *s* is a (closed) sentence'.[21] I take it that 'are equivalent to' here means simply 'have the same truth value as'.

It is clear that if Davidson's project is successfully carried through, comments closely analogous to those I have made on Tarski's procedure will apply to it also. The merit of a Davidsonian 'truth-definition' would lie in the account it would give of the contribution which the internal structure of complex sentences of a natural language makes to their meaning. It is not clear to me that a 'truth-definition' is the best form in which to give this account. I think that some elements in structural meaning (e.g. non-material conditionals) will elude this approach. Even where it succeeds, it gives not a conceptual analysis of 'true' but a factual analysis of a relation between language and the world which holds when and only when a sentence is true-in-the-ordinary-sense. Where 'satisfied' is expanded into the proposed recursive definition, '*s* is satisfied by all functions' is only contingently coextensive (for all sentences in the language in question) with '*s* is true'. A Davidsonian 'truth-definition' may indeed 'yield insight into how, by uttering sentences, we sometimes manage to say what is true' whereas the simple classical definition of truth yields none. But what it yields insight into is the contribution of sentence-structure to meaning; the two kinds of definitions are doing quite different jobs, and it seems to me misleading to call the Tarski-Davidson one a definition of *truth*.

[21] Op. cit., p. 758.

But Davidson, like Tarski, would not be content to say that his account does one job while the classical account does another; he thinks that what the classical account tries to do cannot be done with any accuracy or clarity. His main reasons for thinking this will be considered in § 5 below; but something also emerges from his criticisms of 'the strategy of facts'.

The attempt to construct a correspondence theory of truth where the terms of this relation are, in each case, a sentence and a fact, and the correspondence is thought of as some sort of point-by-point mirroring or picturing of objective elements by sentence-elements, or even a mapping from one to the other, is indeed hopeless. It works all right for the simplest descriptive sentences, but it fails completely for, say, 'Every member is either a parent or a teacher'. But I shall try to show (in § 4) that the simple theory of truth has no need to embrace these absurdities. Davidson, however, argues that the source of all such difficulties is 'the desire to include in the entity to which a true sentence corresponds not only the objects the sentence is "about" (another idea full of trouble) but also whatever it is the sentence says about them'.[22] But is this wrong? Let us see whether the strategy of satisfaction avoids it. If Dolores loves Dagmar, then the two-place predicate-expression '. . . loves . . .' is true of Dolores and Dagmar in that order, it is satisfied by a function that assigns Dolores to the first blank and Dagmar to the second. But what is the closed sentence 'Dolores loves Dagmar' true of? Presumably the universe at the time in question. Since it contains no free variables, no entity-to-variable assignments are relevant; this sentence satisfies all functions trivially, because it is true of things-in-general. It is related to functions much as 'For all trees, lead is malleable' is to trees. Since in going from the open sentence to the closed one we have taken 'Dolores' and 'Dagmar' into the predicate-expression, we no longer need or can have the *objects* Dolores and Dagmar as what it is true of. But it would be silly to say that the entity that this predicate-expression is true of no longer includes Dolores and Dagmar; of course the universe, or things-in-general, at that time includes these people, and if it didn't the closed sentence wouldn't be true of that universe. It is merely that having included their names in the predicate-expression,

[22] Op. cit., p. 759.

we need not refer specifically to them as what it is true of: it is true of everything. But what happens to Dolores and Dagmar when we go from the open sentence to the closed one had happened to the (Fregean) concept *loves*—see § 1 above—already in the open sentence: its name had been put into the predicate-expression, so it no longer needed to be mentioned as part of what that expression was true of. But if the universe hadn't included it (in the appropriate place) that expression would not have been true of Dolores and Dagmar, and the closed sentence would not have been true. The parallelism can be emphasized by using the notion *is sub-true of* which I introduced in § 1, and making our moves in a different order. The (ordered twin) subject-expression 'Dolores . . . Dagmar' is sub-true of the concept (or are-of-a-kind) *loves*. The objective item *loves* must be there for 'Dolores . . . Dagmar' to be sub-true of it. And if we then put the name of this item into the predicate-expression to get our closed sentence again, what this sentence is true of is things-in-general, we need no longer specifically refer to the item *loves*, but the universe the sentence is true of still contains that item, and the sentence would not be true of it if it did not. Of course, none of this is to be taken as suggesting that a (Fregean) concept or is/are-of-a-kind is any sort of object. *Loves* is a universal, and it exists only in such instantiations as *Dolores loves Dagmar*, but it is part of the universe none the less, and the universe would be the poorer without it.

I conclude that there is nothing wrong with including in the entity to which a true sentence corresponds either the objects it is about or how they are said to be, and this is not to be avoided even by taking 'corresponds' as 'is true of'. The mistake in sentence-fact correspondence theories is not this, but trying to find a point-by-point mirroring relationship, whereas many sentence-components have other tasks than to mirror any objective items.

§ 4. *The Austin–Strawson Debate*[23]

Much recent discussion of truth has arisen out of Austin's attempt to present a defensible form of the correspondence

[23] See papers by J. L. Austin, P. F. Strawson, and G. J. Warnock in *Truth*, ed. G. Pitcher, and Strawson, 'Truth: a Reconsideration of Austin's Views', *Philosophical Quarterly*, 15 (1965), 289–301.

theory, and Strawson's criticism of this attempt. Strawson had previously criticized the semantic theory on the grounds discussed above. He had then gone on to present a performatory account of the use of 'is true', equating, for example, 'What the policeman said is true' with 'The policeman made a statement. I confirm it', in which 'I confirm it' is performative, not descriptive: to say this is to confirm the policeman's story, as one could by telling the same story again. Saying that something is true is a device for in effect repeating a story without actually telling it again.[24] On this view, not only is 'is true' applied to statements (what is said) and not to sentences, so that questions of meaning do not arise, but even about statements it has no assertive or descriptive function. 'What the policeman said is true' does indeed re-assert the policeman's story; but it does not make any further assertion about that story.

Austin agrees with Strawson that 'is true' is used in talking about statements, not sentences, and he too therefore rejects the semantic theory, but he maintains that it still has an assertive function: it is a way of describing a 'rather boring yet satisfactory relation between words and the world'.[25] Austin's way of putting this invites the criticism which Strawson had previously applied to the semantic theory, and Strawson naturally turns this against Austin, arguing that what is characteristic and distinctive about Austin's formulation is that it adds to a Ramsey-like account something about words and semantic conventions.[26] Similarly, Ayer says that 'if we strip (Austin's) "definition" of its semantic accretions' we are left with a truism which 'is hardly a vindication of the correspondence theory of truth'.[27] But let us look again at Austin's formula: 'A statement is said to be true when the historic state of affairs to which it is correlated by the demonstrative conventions (the one to which it "refers") is of a type with which the sentence used in making it is correlated by the descriptive conventions.'[28] We could surely read this as follows. We start with a speech-episode, which is the using of a certain sentence in a certain way and

[24] 'Truth', *Philosophy and Analysis*, pp. 272–3.

[25] *Truth*, p. 31.

[26] e.g. *Truth*, pp. 43, 53, 82, and 'Truth, a Reconsideration . . .', 298–300. The reference to Ramsey is to what has been called the Redundancy Theory of Truth in F. P. Ramsey, *The Foundations of Mathematics*, pp. 142–3; the passage is quoted in *Truth*, pp. 16–17.

[27] *The Concept of a Person*, p. 185.

[28] *Truth*, p. 22.

in a certain context, and which thereby is also the making of a certain statement. The sentence, thus used, is correlated by what Austin calls demonstrative conventions with a certain historic situation or state of affairs. The sentence (irrespective of this particular context) is also correlated, separately, by what he calls descriptive conventions with a certain type of situation. And then to say that the statement is true is to say that this historic situation is of this type.

I cannot be sure that this is what Austin intended. To reach this interpretation I have had to substitute 'the sentence, thus used' for Austin's 'it', which refers back to 'statement'; a statement in the sense of what is said cannot be correlated by any conventions with a situation. But if we set this aside as a slip then what I have offered is a possible reading of some of the things Austin said, and while it is open to some criticisms, it is not open to others. In particular it is not open to the charge of saying that to call a statement true is to say that these various semantic conditions are fulfilled, or indeed to talk in any way about semantic conventions. Nor would it be open to the charge that Austin has diverged from his agreement with Strawson against Tarski about what is to be called true, and is here treating 'true' as a description of sentences. It is rather that he has gone too far the other way. Curiously enough, in contrast with what Austin says elsewhere in his paper, on this reading to say that a statement is true is not even to say anything directly about that statement: it is simply to say that a certain historic situation is of a certain type.

This interpretation may be clarified by a diagram.

A statement is made, i.e.,

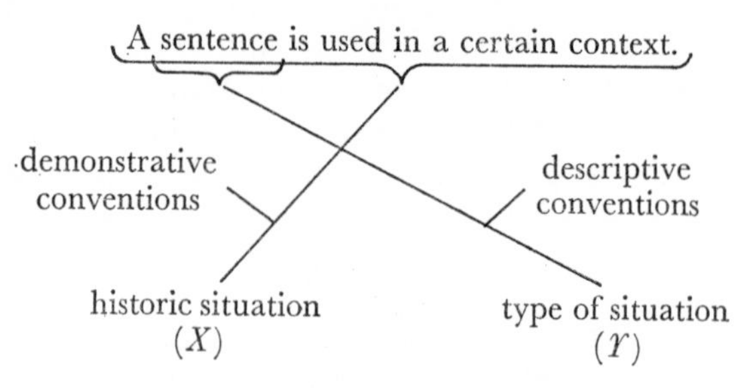

The statement is true $= X$ is of type Y.

The semantic conventions have done their work when their two separate prongs have taken us from the use of the sentence to a situation and to a type of situation: the statement's being true will consist in this situation's being of this type, which of course is a matter of fact, not of convention.

But, as I said, this view is open to other criticisms. We can concede, what seems of no great importance, that the demonstrative and descriptive functions are not performed by two exclusive sets of conventions.[29] The most serious objection is that this account would fail to fulfil what was clearly one of Austin's purposes, to give an account of the statement-that-it-is-true-that-*S* as asserting a relation between the-statement-that-*S* and the world.[30] Another objection is that this view distorts many of the statements to which it is meant to apply by forcing them all into a single mould. It treats them all as being, in a rather special way, subject-predicate statements, where the subject is the indicated historic situation and the predicate is the type of situation. They are all assimilated to the form *This is of that type*, where what 'this' points to may be not, say, a cat, but rather a state of affairs, and the type may be not, say, *black* or *fat* or *hungry*, but rather *of-a-cat-being-on-a-mat-sort*. This restructuring of our statements, though odd, may still be permissible for some restricted range of statements, namely what Strawson has called historical statements. But then there is an objection which Strawson has emphasized, the difficulty of identifying a particular historic situation as being the one with which a certain speech-episode is correlated. We may be unable to find distinct bits of the world with which to correlate what are clearly different statements, or any precise bit of the world with which to correlate a false statement or a true negative statement. If we are to defend the view I have suggested, we must take the subject, this historic situation, as some rather broad and loosely indicated slice of the universe—say a particular room at a particular time—abandoning the picture of a separate and clearly defined 'state of affairs' with which each individual statement corresponds or fails to correspond.

It is worth noting that an analogous but more extreme move is involved in the Tarski-Davidson procedure of going from

[29] Cf. Strawson, *Truth*, pp. 291–4. [30] Ibid., p. 27.

satisfaction to truth. Initially what satisfies a sentential function, what it is true of, is an object or set of objects. But what satisfies a closed sentence, what it is true of, and therefore true, is the universe, all sets of objects. Austin would have said something more closely analogous to this if he had abolished the prong I have labelled 'demonstrative conventions', made the other prong consist of a mixture of demonstrative and descriptive conventions, and made it lead not to a type of situation but to *how things are said to be*. He could then have identified the statement's being true with things-in-general being *how-things-are-said-to-be*.

This might have been better. For Austin's view as I have interpreted it is a predication theory of truth, not a correspondence one—but in the sense of identifying truth not with a *predicate-expression's* being true of some subject (such as we found in the Tarski–Davidson procedure) but rather with a Fregean concept's belonging to a certain object. For Austin's type of situation is closer to a Fregean concept, an objective predicate or *is-of-a-kind*, than to a predicate-expression. The result, however, is that truth has been identified with a wholly non-linguistic, non-semantic relation, a situation's being of a certain type. What has this to do with the relation between a certain statement and the world, let alone between *words* and the world? Suppose we answer that the statement in question is, after the proposed restructuring, the statement that this situation is of this type, and that is why this situation's being of this type makes it true. Admittedly this is what *makes* the statement true. But if we identify its *being* true with this situation's being of this type we collapse the statement that it is true that S into the statement that S: we are back at Strawson's re-affirmation theory and have failed to allow for the distinctive, extra, assertion for which Austin was trying to provide. To say, distinctively, that the statement is true is not merely to say that this situation is of this type, but to note the connection between this and what is stated. To say that the statement is true is not merely to say that X is of type Y, but to say that *as was stated* X is of type Y. We need some such comparison to preserve the force of 'true'. Without it, we should have an extreme Ramseyan thesis, such as Strawson did endorse in 1949, but which would deny what he described in 1964 as the undisputed thesis that 'someone who says that a certain statement is true thereby makes a statement about a statement'.[31]

If we are to avoid this extreme Ramseyism, we must revise the interpretation I offered and say something more like this:

> Where a statement can be restructured as the statement that a certain (loosely indicated) situation is of a certain type, to say that the statement is true is to say that, as is stated, this situation is of this type.

But having got so far, why should we retain the restriction to historical statements? Why not generalize to something like the following:

> For all *p*, to say that the-statement-that-*p* is true is to say that, as is stated, *p*.

However, there is still something queer about this formula. It equates saying that the-statement-that-*p* is true with making two claims, namely that *p* and that this is as is stated. This account combines an element of re-affirmation with an element of comparison. But once we have distinguished these elements, we see that it is that of comparison that is vital for the ascription of truth. To say that the-statement-that-*p* is true is to say that things are as they are stated to be; it is not essentially to re-affirm that *p*. Of course, if things are as they are stated to be, then *p*; and if we know that the statement is the-statement-that-*p* we cannot say it is true without implicitly asserting that *p*. But still this is not what we are primarily doing when we say that this statement is true. This comes out clearly if someone in whose encyclopaedic knowledge and unswerving veracity I have complete confidence has made a statement, but I have not heard, or perhaps not understood, his utterance. I can still say not merely that his statement, that is, speech-act, was veridical, but also that his statement in the other sense, what he stated, the proposition he asserted, is true. I can say 'Whatever he said, it's true'; and surely I am then saying merely that things are as they were said by him to be. I am not re-affirming his statement, since I am not in a position to do so. No doubt I am prepared to re-affirm it; but this readiness is not arbitrary, it does not stand on its own; it results from what I first believe about whatever he has said.

[31] *Truth*, p. 68.

Thus we come back to the simple thesis: *To say that a statement is true is to say that things are as, in it, they are stated to be.*

We have reached this formulation, or rather returned to it, by criticism and correction of Austin's formula. We can reach it also from things that Strawson says or concedes in his later contributions to the debate. He mentions this possible ruling: 'A historical statement is true if there obtains the historical state of affairs which, in the making of the statement, is stated to obtain.' Strawson sets aside the conjunction of this with a remark about semantical conventions as not being what is desired by a defender of Austin; but he regards the part quoted here as uncontroversial.[32] It remains uncontroversial if we generalize beyond the sphere of historical statements, and say:

> A statement is true if there obtains whatever in the making of the statement is stated to obtain.

This says, almost trivially, *in what circumstances* a statement is true. But surely we can take one small step further and use this as an account of *what we are saying* when we say that a statement is true:

> To say that a statement is true is to say that whatever in the making of the statement is stated to obtain does obtain.

And this is equivalent to the simple thesis above. But what have we now? An account of the use of 'true' which is concerned not at all with re-affirmation but only with comparison. It locates truth in a certain relation, not between words and the world but between what is stated (or how things are said to be, or a proposition), and the world or some part of it. Consequently it incorporates what Strawson says or concedes in replying to Warnock, that (in predications of 'is true') 'something is said both about how things are in the world and . . . about a statement'.[33] But only very little is said about either, indeed nothing is said about either on its own: all that is said is that things are in the world *as* in the statement they are stated to be. If the content of the statement is given, then by implication how things are in the world is specified; but this specifying is not the work of the predication 'is true'.

This formulation, then, denies that the primary function of

[32] 'Truth, a Reconsideration . . .', 298–9. [33] *Truth,* p. 81.

'is true' is either wholly or partly to re-affirm the statement that is said to be true. A pure re-affirmation account, such as Strawson gave in 1949, would say that the primary function of '*S* is true', where *S* is the statement that *p*, is to re-affirm that *p*; a partial re-affirmation account would say that the function of '*S* is true' is to say that as is stated in *S*, *p*. But what I have offered, in contrast with both of these, is a pure comparison account: the primary function of '*S* is true' is simply to say that things are as is stated in *S*.

Rejecting both re-affirmation accounts, we must explain why they have seemed plausible. It is not difficult to do so. Their plausibility may arise partly from attention to cases where the statement which is said to be true is identified directly, as a specific *that-p*; in these cases there is an implicit or consequential re-affirmation that *p*. Again, the truth-conditions of '*S* is true' will coincide with those of *S*, which may lead us to infer, wrongly, that these statements must have the same meaning. The conversational point of saying that *S* is true is often, though not always, to re-affirm *S*, and it is easy to confuse the conversational point of a remark with its basic meaning. But over and above all of these there is a historical, etymological explanation. The original meaning of 'true' connects it with trust, with being trustworthy, loyal, and so on. A true statement was originally very like a true friend. Of course, 'This is trustworthy' has still a descriptive basic meaning. But it lends itself very readily to a performative use. 'That is true' thus naturally has the illocutionary force of 'I vouch for that'; 'is true' can function like an assertion-operator, and in doing so it preserves a stronger link with its past than when it bears the simple sense which is now central.[34] Nevertheless the simple sense which I have been trying to clarify is now central, and it is, as I have indicated in § 3, this simple sense which ensures that meaning is related to conditions for truth in the way that the Tarski–Davidson procedure exploits.

Ramsey raised the question whether the words 'true' and 'false' are eliminable, even where the statements to which they are applied are not given explicitly.[35] I think they are, but this does not commit me to a re-affirmation account rather than

[34] J. R. Lucas, *The Concept of Probability*, pp. 1–2.
[35] *The Foundations of Mathematics*, p. 143.

to one of comparison. A comparison account might be summed up by this equation:

(1) 'S is true' $=(\exists x)((S$ is the statement that $x)$ & $x)$

The word 'true' is here eliminated, but truth is not eliminated but displayed: the relation in which it consists is made clear. What I have called the extreme Ramseyan or pure re-affirmation account might, contrastingly, be summed up thus:

(2) $(x)((S$ is the statement that $x) \supset$ ('S is true' $= x))$.

And this would eliminate truth by equating the statement that S is true with the statement that x, i.e. with S itself.

But it is clear that (1) and (2) are not equivalent. I maintain that (2) as it stands, with '=' as a sign of analysis or meaning, is mistaken. But the somewhat similar formula,

(3) $(x)((S$ is the statement that $x) \supset ((S$ is true$) \equiv x))$,

where '$\equiv$' is the sign for material equivalence, is correct, and is derivable from (1). We may be tempted towards a re-affirmation account by informal thinking which amounts to a failure to distinguish (2) from (3).

I have introduced these formulae to help to explain why re-affirmation accounts seem plausible. I do not intend that the bare formula (1), without all the discussion throughout this chapter, should be taken as an adequate explication of truth. Nevertheless I think that such a formalization is correct. Some philosophers have doubts about formulae of this kind, in particular about the intended range of the variable 'x', which occurs both after 'that' and on its own after '&'. I shall try to resolve these doubts in § 5. Other philosophers may object that any analysis of which (1) is a summary explains truth in terms of statements, and that it will not then be possible, without circularity, to define statements—as opposed, say, to commands or questions—in any way that is parasitic upon truth and falsehood. There is indeed a risk of circularity here, but we simply have to break into the circle at some point. For example, we are familiar with asserting, commanding, and inquiring; we can distinguish them and we have learned names for these procedures in association with the procedures as we have encountered them and engaged in them. Since this is so, we can define a statement as something assertible, as the content of an assertion, in the way defended in § 1, and there will then be no circularity in an account of truth that makes use of this notion of statements.

I have suggested that the formulation, *To say that a statement is true is to say that things are as, in it, they are stated to be*, is uncontroversial. But it may not be. Strawson's 1950 paper includes the dictum that 'The trouble with correspondence theories of truth . . . is the misrepresentation of "correspondence between statement and fact" *as a relation, of any kind, between events or things or groups of things*',[36] and this is backed up by an argument to show that facts, situations, states of affairs, etc., are not parts of the world. This argument, however, does not disprove the possibility of a relational view. The association of facts with that-clauses shows only that the parts of the world in question are most naturally specified in a way that is parasitic upon statements, and perhaps that they cannot be specified exactly in any other way; but for all that something quite objective can be indicated by the vague phrase 'how things are in the world' or by at least some uses of such words and phrases as 'fact', 'situation', 'state of affairs', and so on. Strawson's 1950 arguments are intended to bear against *any* relational view, and by implication therefore against even the very modest thesis I have presented; but I would appeal against them to what Strawson regarded as uncontroversial in 1965; all that the earlier arguments really tell against is the point-by-point sort of correspondence theory which requires a precise fact to set alongside each statement or sentence, with one-one correlations between their components.

Alternatively, it might be argued that although the proposed formulation is uncontroversial, the description of it as relational is not so, not because of difficulties about the world end of the relation, but because of difficulties about the statement end of it. Whereas statements as speech-episodes may be entities real and independent enough to serve as terms in a relation, statements as what is stated, or how things are said to be, are not. Admittedly there is a risk that in calling this a relational theory we may suggest that the statement end of the relation is more of an independent entity than it is, and this, combined with the use of the word 'statement', may tempt us to take as the

[36] *Truth*, p. 40.

unworldly term of the relation the statement in the other sense, the speech-episode. This temptation would be reduced if we spoke of propositions rather than of statements, but propositions are liable to be reified in another way. In any case, the temptation can be resisted. If we do resist it, I see no objection to speaking about a relation between a proposition or statement or how things are said to be and how things are. As Austin says, there is a range of possible relations here, including but going beyond things being as stated (truth) or not being as stated (falsehood).[37] Admittedly what is inept is a speech-episode rather than a proposition, but the latter can be exaggerated, precise, and so on. Some of these characterizations point to triadic or polyadic relations involving the hearer's purposes, previous knowledge, etc., as well as what is stated and how things are, but the last two items at least are involved.

Both these objections to this attempt at a relational account may be combined in the complaint that the two items which it proposes to relate are both too shadowy, that each of them in fact is modelled upon sentences. It is because statement and fact are each a mere reflection of a sentence that they can appear to correspond so closely, and yet so uninterestingly. We can concede that it is only by using some sentence (or that-clause) that we can specify a statement, a how-things-are-said- (or-believed) -to-be, with any precision, and similarly that it is only by using some sentence (or that-clause) that we can specify a fact. But there is no reason for identifying something with the way in which it is specified or introduced. We cannot deny that a how-things-are-said-to-be is the content of certain speech-episodes, that the very same content may occur, as what is believed, when someone believes something. And equally the fact that is specified by some sentence, a certain how-things-are, occurs quite independently of the sentence that is used to introduce it.

Essentially the same objection can be put in another way. Prior has said that 'facts and true propositions alike are mere "logical constructions" . . ., and . . . they are the *same* "logical constructions" (to have "true propositions" *and* "facts" is to have *too many* logical constructions)'.[38] When Prior explains

[37] *Truth*, p. 28. Cf. Strawson, *Truth*, pp. 49–50.
[38] A. N. Prior, *Objects of Thought*, pp. 5, 12.

why propositions are logical constructions, he does so by expanding this into the assertion that 'sentences that are ostensibly about propositions . . . are not in reality about propositions but about something else'. 'Part of what it says may be, e.g. that the sentence "The proposition that the sun is hot is true" means no more and no less than the sentence "The sun is hot".' But this is just the pure re-affirmation theory which we have already found reason to reject. Another point he makes is that '*X* believes' (or 'fears', or 'hopes', or 'desires', and so on) 'that so-and-so' is better understood as divided thus:

'*X* believes that/so-and-so' than thus:
'*X* believes/that so-and-so'.

The latter division encourages us to take believing as a relation between *X* and a pseudo-entity, a proposition; the former dispenses with these pseudo-entities, and makes 'so-and-so' merely a sentence: '. . . believes that . . .' is an expression which forms sentences out of two components, a name and a sentence. But Prior insists that this does not commit us to saying that what we believe, or hope, or fear, etc., are sentences.[39] This seems to me correct as a way of denying that propositions are entities, of eliminating apparent names of the form 'that so-and-so'. But this does not prevent us from picking out how things are said to be with a view to comparing this with how things are. If this is all that is meant by calling propositions logical constructions, well and good. But then the crucial point is Prior's other contention, that facts are the same logical constructions as true propositions. But why are facts logical constructions? Is talking about the fact that the sun is hot really talking about something else? Admittedly it is talking about the sun, but not just that: rather about the sun's being hot. I do not see how we can deny that a particular aspect of reality is picked out by the sentence 'The sun is hot', and that this quite objective feature is what we call the fact. As Ramsey says,[40] 'The fact that *a* has *R* to *b* exists' is no different from '*a* has *R* to *b*'; but what this last sentence reports is something quite real and concrete, and the word 'fact' merely enables us to talk generally about this sort of thing. Of course if facts were identical with true propositions, and propositions (true or false) were logical constructions, facts

[39] Op. cit., pp. 14–20. [40] Op. cit., p. 143.

would have to be logical constructions too. But what are the grounds for this identification? Merely the near-equivalence of such pairs of sentences as 'It is a true proposition that *p*' and 'It is a fact that *p*'. But in speaking of 'the proposition that *p*' we are speaking of the content of some belief-state, assertive speech-episode, or the like, whereas in speaking of 'the fact that *p*' we are speaking about some aspect of the world. The same sentence '*p*' is used to pick them out, but what it picks out is very different in the two cases. But because a proposition's being true consists in things being as they are said to be, wherever there is a true proposition that *p* there must also be a fact that *p*. Does the converse hold? It might be best to keep 'proposition', 'statement', etc., for the contents of *actual* believings, sayings, and so on, and then of course there will be many facts with no corresponding true propositions, though not of course facts that we are in a position to specify. Alternatively, it is very natural to extend the terms 'proposition' and 'statement' to cover what might be, as well as what is, said, believed, etc.: and then there will be a true proposition for every fact. 'It is a true proposition that *p*' and 'It is a fact that *p*' will always be materially equivalent. But still not synonymous. Of course, one must concede that these are slippery terms. There is a way of using 'fact' so that it is synonymous with 'true proposition', so that it shifts from pointing to an aspect of the world and points to the content of an actual or possible belief or assertion. Similarly there is a way of using 'proposition' so that true propositions are taken as parts of the world, and then we are led to postulate imaginary false propositions as their analogues. But both these moves can and should be avoided: there are two sorts of item that can be distinguished, and truth can consist in a relation between them.

Even if this relational account is accepted, it may be questioned whether it can be called a form of the correspondence theory. Admittedly the term 'correspondence' is less appropriate here: words or sentences or speech-episodes or states (or acts) of believing might be held to *correspond* in some way that could be spelled out in detail, some specific mapping from one realm to the other might be set up; but the relation between how things are said to be and how they are (when all goes well) is too close and intimate to call for any specific mapping, and therefore too close to be called correspondence.

This point might be conceded, or rather welcomed. Although I suggested (in § 3) that the correspondence theory comes in as an opponent of such sceptical or otherwise evasive theories as the coherence theory and the pragmatist theory, it is itself a sceptical theory of a milder kind. If the best we could achieve was that our statements should somehow correspond to what is there, we should still be falling short of having things just as we state them to be. If anyone is dissatisfied with my various formulations on the ground that I still need to explain and expand the 'as'—for example in 'things are as they are said to be'—then he has not grasped the point of my phrase 'the simple theory of truth'. Any expansion would loosen the connection. The propositional content of an utterance or belief can be just how things are—though never more than part of how things are—though of course a proposition or statement, existing only as the content of an assertion, belief, etc., cannot be numerically identical with any part of the world.

Admittedly this *is* a simple theory. I have not said positively in the end any more than it seemed obvious that we should say at the start. If this discussion has any merit, it is in the avoidance of traps into which the example of various distinguished thinkers shows that it is all too easy to fall. To give a correct account of the central ordinary sense of 'true' is like walking along a very narrow path with an abyss on either side. To do justice to the notion of comparison, of some relation between statement and reality, one is tempted to bring in words or sentences or concrete belief-states, and so to substitute for the conceptual analysis of truth the factual analysis of the surroundings in which issues of truth arise, to give an account of something that is at best coextensive with truth, an account whose interest lies in the contribution it makes to the theory of meaning (or, perhaps, to the psychology of belief). But if one avoids this it is easy to lose the central ordinary sense of 'true' and to substitute for it some illocutionary force which assertions of truth very naturally have on account both of their conversational function and of the ancestry of the word 'true'.

Although the account I have given is too simple and obvious to be of much positive interest in itself, it does enable us to raise some further issues more sharply than we could without it. In particular, it is of importance elsewhere in philosophy to distinguish what are from what are not issues of truth and falsehood. The problem will now be to decide how widely we can apply the description we have been analysing. I have hinted (in § 2) at some arguments that can be used against the extreme scepticism which denies that we can ever achieve simple truth; but it may still be that we cannot achieve it in as many areas as we think we can. For example, I shall argue (in Chapter 3) that conditionals, other than material conditionals, in their primary employment are more like arguments than like statements about how things are, and that some of them are not capable of being simply true. It is more plausible to give a performative account of the basic, central, meaning of 'is probable' than of the central sense of 'is true'; but as we shall see in Chapter 5 probability judgements are of a number of different kinds, and some are and some are not capable of simple truth. It may well be that moral judgements and judgements of value generally are not capable of simple truth, and it has been suggested that this holds for all open universal judgements too. But these become sharp issues only if we are using the realist conception of simple truth; without it we have nothing definite to contrast with cases where although this conception does not apply, our judgements are not purely arbitrary or subjective, but may well be reasonable, warranted, defensible, and so on. For example, Tarski's 'semantic' requirement that there should be the same truth-conditions for '*p*' and for ' "*p*" is true' would not in itself make these sharp issues—there are presumably the same defensibility conditions for '*p*' and for ' "*p*" is defensible'—and similarly if we were using a pure re-affirmation theory of truth it would be absurd to suggest that conditionals cannot be true.

§ 5. *Problems of Symbolic Formulation*

In Tarski's elementary examples we find sentences of the form ' "*p*" is true if and only if *p*', where the first occurrence of '*p*' is as a sentence in the object language, the second as a translation of this into the metalanguage: it can be visibly the same sentence if the object language is a proper part of the metalanguage. A similar and yet different repetition occurs in, e.g., Prior's formulation of '*X*'s belief that *p* is true' as '*X* believes that *p* and (it is the case that) *p*'.[41] As Prior indicates, we can leave out 'it is the case that'. Here '*p*' occurs both times in the object language: the sentence in question is twice used and never mentioned;[42] but it is used first as a component in the larger sentence '*X* believes that *p*' and secondly as a sentence on its own, merely conjoined with the other. I would agree with Prior's insistence that there is nothing improper or mysterious in this double use of the same sentence on the same language level, either operated on by a sentence-forming operator on sentences, '*X* believes that . . .' or not. Admittedly this type of construction makes it possible to formulate logical paradoxes of a peculiar kind, which Prior has discussed and which I shall consider in Chapter 6; but this possibility does not show that this type of construction is itself objectionable.

Once this is allowed, it becomes easy to formulate '*X* believes truly' (or 'falsely') 'that *p*' and in so doing to eliminate the terms 'truly' and 'falsely'. But to formulate in the same style other things that we may want to say about truth and falsehood we shall need to do more than this, we shall need to quantify over such items as this '*p*' in both its occurrences. Thus I have offered (in § 4 above)

$$(\exists x)((S \text{ is the statement that } x) \ \& x)$$

as a formulation of '(the statement) *S* is true'. Similarly, Ramsey formulates 'He is always right' as 'For all *p*, if he asserts *p*, *p* is true' but notes that the final 'is true' can be dropped.[43] Davidson, in sketching (before criticizing) the 'redundancy theory' translates 'The Pythagorean theorem is true' as

$$(p) \text{ (the statement that } p = \text{the Pythagorean theorem} \rightarrow p),$$

'Nothing Aristotle said is true' as

$$—(\exists p) \text{ (Aristotle said that } p.\ p),$$

and gives as a principle which would explain truth as a genuine predicate of statements

$$(p) \text{ (the statement that } p \text{ is true} \leftrightarrow p).^{44}$$

[41] *Objects of Thought*, p. 22.
[42] Op. cit., p. 99.
[43] Op. cit., p. 143.
[44] 'True to the Facts', pp. 749–50.

Similarly again, C. J. F. Williams argues that the predicate (of statements) 'is true' can be symbolized by

$$(\exists p)[(\ldots \text{ states that } p) \,\&p)$$

where the gap indicates where the name of the subject statement is to be inserted.[45]

Many philosophers are suspicious of such quantifications. Davidson, for example, goes on to say that such formulations would explain truth if we could understand them. But, he asks, do we? 'The trouble is in the variables. Since the variables replace sentences both as they feature after words like "Aristotle said that" and in truth-functional contexts, the range of the variables must be entities that sentences may be construed as naming in both such uses. But there are very strong reasons, as Frege pointed out, for supposing that if sentences . . . name anything, then all true sentences name the same thing.' And he hints that 'Paradox may also be a problem for Ramsey's recursive project'.[46] Similarly, Quine says that any higher-order quantification brings with it new ontological commitments.[47]

But these difficulties are all chimerical. The problem of paradox will be discussed in Chapter 6. The others have, I think, been effectively disposed of by Prior.[48] What a variable 'stands for' in one sense is simply what it keeps a place for, and the variables in the above formulae keep places for sentences. There is no need to suppose that they 'stand for' anything at all in the other sense of designating some objects. If each item for which 'p' keeps a place does not designate an object—and the job of sentences is not to designate objects—then 'p' doesn't stand for anything in this second sense. If sentences denoted anything, it might have to be the True or the False; but they don't.[49] Briefly, there is no reason why these quantified formulae should not function exactly as they are intended to, namely as compendious ways of speaking about the sorts of things we get when their variables are replaced by sentences. We should indeed be in difficulties if we had to find a single category of entities which can occur both on their own, as parts of the world,

[45] C. J. F. Williams, 'What does "x is true" say about x?', *Analysis*, 29 (1968–9), 113–24. This article defends the view that 'is true' is a genuine predicate of statements on grounds very close to those that I have offered.

[46] 'True to the Facts', pp. 750–1.

[47] *From a Logical Point of View*, pp. 112 ff.

[48] *Objects of Thought*, ch. 3.

[49] Op. cit., pp. 50–2.

and as the contents of beliefs, assertions, and so on. But these quantifications do not require this. The repeated variable merely indicates that in each instantiation the same sentence will be used twice, once to pick out a statement, how things are said to be, and once to pick out a fact, how things are. It is the sameness of sense between these two occurrences that we are using, not of reference on the interpretation which gives to all true sentences the reference 'the True'—though if the sentence includes names or indexical terms we must include what it achieves thereby as part of its sense. These quantifications apply to sentences rather than to any entities that sentences might be held to designate, but to sentences used, not mentioned: we are not to read '$(\exists p)$', for instance, as 'There is a sentence "*p*" . . .'. And we should not let anyone tell us that we cannot read these quantifiers and variables in the intended way; the symbols are our instruments, not our masters.

Are we here using what Quine calls *substitutional* quantification?[50] As he defines it, 'An existential substitutional quantification is counted as true if and only if there is an expression which, when substituted for the variable, makes the open sentence after the quantifier come out true', and there is a corresponding rule for universal quantification. This would be sufficient for most of our purposes, since in general there will be expressions for the statements which we want to call true. But, it might be objected, there are some purposes for which it would not be sufficient: we may want to say that there are non-denumerably many truths, but only a denumerable infinity of expressions. If so, there must be true statements which are not expressed by any sentences, and the truth of one of these could not be formulated by '$(\exists x)((S$ is the statement that $x)$ $\& x)$' read as a substitutional quantification. However, this difficulty might be met in more than one way. First, we might speak about possible expressions, some of them infinitely long: there will be non-denumerably many of these. For example, there will be non-denumerably many truths of the form '*d* is greater than 0 and less than 1', where '*d*' is a non-terminating decimal; but the truth of each of these is adequately expressed by '(*S* is the statement that . . . is greater than 0 and less than 1) & . . . is greater than 0 and less than 1' with the understanding that the

[50] *Ontological Relativity*, pp. 104–7.

same non-terminating decimal is to occur in both spaces. Secondly, what comes to much the same thing, the expression used in a substitutional quantification can itself contain a variable governed by a 'classical' quantifier, and can so cover non-denumerably many truths. For example, since every subset of the natural numbers is countable we may want to say that (y) (if y is a subset of the natural numbers then the statement that y is countable is true). Inserting into this the proposed formulation for 'true' we get:

> (y) (If y is a subset of the natural numbers then $(\exists x)$ ((the statement that y is countable is the statement that x) & x)).

In this, '$(\exists x)$ . . .' can be a substitutional quantification; its truth is ensured by the correctness of the result of substituting 'y is countable' for 'x', giving '(the statement that y is countable is the statement that y is countable) & y is countable', which will be true wherever y is a subset of the natural numbers. Thus although our original statement covered non-denumerably many truths—since there are non-denumerably many such subsets—the use of 'true' in it could be expressed by a substitutional quantification.

But even if there were uses of 'true' for which these methods failed this would not be a disaster; at most we should have to admit that these uses represented some extension of the basic sense of 'true' which we have explained. Although Quine seems to suggest that the only alternative to substitutional quantification is classical or *objectual* quantification which carries with it ontological commitments, the ontology is of a very thin-blooded sort. The crucial point is that if a quantifier governs several occurrences of a variable, the quantification can be understood only in so far as we understand what will constitute the *same* item instantiating these various occurrences. In substitutional quantification, the necessary identity is secured by the identity of an *expression*; in any other sort of quantification it will have to be provided for in some other way. The commitment to 'ontology' is merely a commitment to identification. Even if we went beyond substitutional quantification we should not be committed to entities, as suggested above, that occur both on their own and as the contents of beliefs and assertions.

But these are further problems. I suggest that the formulations can be understood so far as we need them. Consequently they, and the informal discussion which they summarize, do after all explain the central ordinary sense of 'true', which neither collapses into redundancy or into a purely performative function nor needs to be replaced by coherence, or usefulness, or a psychological concept, or an analysis, however fruitful, of structural meaning.

3

CONDITIONALS

§ 1. *Introduction—Rival Accounts of Conditionals*

CONDITIONAL sentences and if-sentences—which may not be exactly the same thing—play a large part in our discourse. 'If we join the Common Market . . .', 'If I take that pawn . . .', and similar remarks enter into the consideration of choices and policies; 'If he hadn't operated . . .' and the like into judgements about past decisions; 'If it's a pre-Roman site . . .', 'If it's appendicitis . . .' figure in processes of discovery and diagnosis. Contrary-to-fact conditionals not only figure in retrospective judgements; they seem to be part of the analysis of causal statements—to say that the spark caused the explosion seems to mean at least that if there had been no spark this explosion would not have occurred; but equally 'I can if I choose' and 'He could have if he had chosen' are classic assertions of apparently contra-causal freedom. 'Ifs' are used to make general statements—'If a number ends in 5 it's divisible by 5'—and questions and commands are often conditional. Dispositional statements, such as those which say that someone is irritable or weak-willed or that something is fragile or insecure or has a certain rate of radioactive decay or even a certain mass also seem to entail conditionals (e.g. 'If it were dropped (suitably) it would break') and perhaps involve conditionals in their analysis; and many of the ordinary defining properties of things are dispositional.

But despite their commonness and the apparent ease with which we handle them, conditionals are in some respects obscure. What exactly do they mean? Logicians are constantly complaining that we do not know how to construe them or precisely what their truth conditions are.[1] Contrary-to-fact conditionals—'counterfactuals' for short—have aroused the

[1] D. Davidson, 'Truth and Meaning', *Synthese*, 17 (1967), 304–21, esp. 321.

most anxiety and dispute.[2] They are important because they mark a difference between statements of natural or causal law and merely accidental generalizations. Law statements entail or sustain counterfactuals whereas accidental generalizations do not. If it just happens to be the case that everyone in this room understands Italian, it does not follow that if Mr. Chou En-Lai had been here he would have understood Italian, but if there is a causal law which connects being in this room with understanding Italian then this counterfactual does follow. A law-governed counterfactual extends the scope of a causal law beyond the actual facts and makes it cover possibilities as well. This, it seems, compels us to reject a Humean or regularity theory of causation, and to admit that causal laws involve connections that are not reducible to concomitances and sequences. But such connections would constitute a difficulty for empiricism. How could we discover them? What would they even *be*—for it will not do to define a connection simply as *that which* sustains a counterfactual? Or, looking at the same problem from another side, what would make a counterfactual true?

But it seems plain that counterfactuals differ from other non-material conditionals only by adding something extra. 'If he had come he would have enjoyed himself' differs from the corresponding open conditional 'If he came he enjoyed himself' only by suggesting, fairly strongly, that he did not come and, less strongly, that he did not enjoy himself. If we had a satisfactory account of the open conditional, we should obtain a satisfactory analysis of the counterfactual by merely adding these hints or suggestions. If we cannot find an interpretation that satisfies us of the counterfactual it is very likely that we

[2] Cf. R. M. Chisholm, 'The Contrary-to-Fact Conditional', *Mind*, 55 (1946), reprinted in *Readings in Philosophical Analysis*, ed. Feigl and Sellars, pp. 482–97; N. Goodman, 'The Problem of Counterfactual Conditionals', *Journal of Philosophy*, 44 (1947), reprinted in *Fact, Fiction, and Forecast*; S. Hampshire, 'Subjunctive Conditionals', *Analysis*, 9 (1948), reprinted in *Philosophy and Analysis*, ed. Margaret Macdonald, pp. 204–10; E. Nagel, *The Structure of Science*, ch. 4; W. Kneale, 'Natural Laws and Contrary-to-Fact Conditionals', *Analysis*, 10 (1950), reprinted in *Philosophy and Analysis*, pp. 226–31; G. H. von Wright, *Logical Studies*, pp. 127–65; N. Rescher, 'Belief-contravening Suppositions', *Philosophical Review*, 70 (1961), 176–96, and *Hypothetical Reasoning*; R. S. Walters, 'The Problem of Counterfactuals', *Australasian Journal of Philosophy*, 39 (1961), 30–46; J. L. Mackie, 'Counterfactuals and Causal Laws', *Analytical Philosophy*, ed. R. J. Butler, pp. 66–80; R. C. Stalnaker, 'A Theory of Conditionals', *Studies in Logical Theory* (American Philosophical Quarterly Monograph), pp. 98–112.

have not found an adequate interpretation of the open conditional either.

At least six kinds of interpretation of conditional statements have been suggested.[3] First, there is the view that if-sentences in ordinary speech mean the same as the logical formula '$P \supset Q$' (or '$P \rightarrow Q$') of the propositional calculus, that is, that 'If *P*, *Q*' is equivalent to 'Either not-*P* or *Q*' and to 'Not both *P* and not-*Q*'; the meaning of 'If *P*, *Q*' is given by its truth-conditions as set out in the standard truth table for '$\supset$', namely that it is true whenever '*Q*' is true and whenever '*P*' is false, and is false only when '*P*' is true but '*Q*' false. On this view, ordinary if-sentences are material conditionals after all. This is a paradoxical view, because the problem of construing if-sentences in ordinary speech usually arises because they appear not to be material conditionals. It is the problem of elucidating *non-material* conditionals: but on this view there is no such thing. Of course, material conditionals are not paradoxical in themselves; '$P \supset Q$' asserts a perfectly clear (though incompletely specified) state of affairs, and the so-called paradoxes of material implication, that '$P \supset Q$' is made true by the falsity of '*P*' alone or by the truth of '*Q*' alone, are paradoxical only when '$P \supset Q$' is read as 'If *P*, *Q*'. For the sake of clarity and precision, it would be nice if ordinary if-sentences could be interpreted as material conditionals. I shall discuss why it is paradoxical so to treat them, and whether this paradox can be resolved.

A second account of conditionals is that their meaning is

[3] These 'kinds of interpretation' are 'ideal types' rather than accounts that can be assigned to individual authors, who often adopt what looks like a mixture of two or more of these kinds of interpretation. The materialist account is defended by H. P. Grice in his (unpublished) William James Lectures; a 'logical powers' account is given by G. Ryle in ' "If", "So", and "Because" ', *Philosophical Analysis*, ed. M. Black, pp. 323–40, and a similar account is used by R. M. Hare, 'Meaning and Speech Acts', *Philosophical Review*, 79 (1970), 3–24, esp. 16; P. F. Strawson, *Introduction to Logical Theory*, pp. 35–8 and 82–90 gives mainly a consequentialist account, with hints of a metalinguistic one and a small concession to logical powers; von Wright, in speaking of conditional assertion, seems to be getting towards what I call below (§ 9) a suppositional account, but he also gives a partly materialist account ('to assert *Q* on the condition *P* is to assert that *P* materially implies *Q* without asserting or denying the antecedent or the consequent of the material implication') and a partly consequentialist one; Chisholm and Goodman offer metalinguistic accounts, while bringing out the difficulties of this approach; condensed argument accounts are given by Walters and in my article mentioned in n. 2 above; possible worlds accounts by Stalnaker and (in an as yet unpublished work) by D. K. Lewis.

given by their logical powers. Logicians for over two thousand years have recognized the validity of the argument-forms *modus ponendo ponens*—'If *P*, *Q*; and *P*; therefore *Q*'—and *modus tollendo tollens*—'If *P*, *Q*; and not-*Q*; therefore not-*P*'—and it is plausible to suggest that at least the core of the meaning of 'If *P*, *Q*' is that the conjunction of it with '*P*' entails '*Q*' and the conjunction of it with 'not-*Q*' entails 'not-*P*'. To say that the *whole* meaning of 'If *P*, *Q*' is so given would, however, be surprising. This approach declines to give any explicit meaning to 'If *P*, *Q*' on its own; it says that this if-sentence is either just a something-or-other which enters into these entailments, or the giving of a licence to draw these inferences.

A third account is the consequentialist one: to say 'If *P*, *Q*' is to say that the (possible) state of affairs described by '*Q*' is a consequence of some sort of that described by '*P*'—or, briefly, that *Q* is a consequence of *P*. What sort of consequence differs from case to case. 'If some man loves every woman then every woman is loved by some man' presumably says that every woman's being loved by some man is a logical consequence of some man's loving every woman. But 'If everyone is somebody then no one's anybody' says merely that the latter state of affairs is a social or political consequence of the former. And other if-sentences will similarly present causal, moral, legal, religious, and magical consequences. This view formulates and builds upon our most natural objection to the material conditional interpretation. 'If Senator Mike Mansfield is a Republican then Adelaide is south of Melbourne' and 'If pumice floats then Senator Mike Mansfield is a Democrat', taken as material conditionals, would both be true, simply because Senator Mansfield is a Democrat and not a Republican, but we are naturally inclined to say that neither is an acceptable conditional because in each case the consequent has nothing whatever to do with the antecedent, because the one is in no way a consequence of the other. However, this point must be made with caution. It is not difficult to construct a situation in which both these conditionals are acceptable. Suppose we have a general knowledge quiz contest in which competitors are presented with groups of statements, having been told that each group consists either of true statements only or of false statements only, and one group includes 'Senator Mike Mansfield is a

Republican' and 'Adelaide is south of Melbourne' while another group includes 'Senator Mike Mansfield is a Democrat' and 'Pumice floats'. A competitor in this contest, pondering what to say, might very reasonably use both these conditional sentences. But this, the consequentialist will say, is because in this special situation a pair of consequential relations has been set up; given the background information, Adelaide's being south of Melbourne is a consequence of Senator Mansfield's being a Republican. If his account is to cover all acceptable if-sentences, he must take 'being a consequence' in a wide enough sense to include such artificially constructed cognitive consequences; and he must not dismiss as absurd any if-sentence, however bizarre, however apparently unrelated its antecedent and consequent may be, but content himself with saying that it is prima facie unacceptable, that it doesn't make sense as an ordinary if-sentence until some consequential connection is filled in.

This leads very readily to a fourth, metalinguistic, type of account. We can understand even a bizarre if-sentence, even before any consequential connection has been filled in, because we take it as saying that *some* consequential connection holds between *P* and *Q*. 'If *P*, *Q*' is on this view equivalent to a metalinguistic statement, 'Either "*P*" entails "*Q*" or there is some statement or set of statements "*S*" which is true, and cotenable with "*P*", and the conjunction of which with "*P*" entails "*Q*".' In the example given it is easy to find such a set of statements which along with 'Senator Mansfield is a Republican' entails 'Adelaide is south of Melbourne'. The qualification that '*S*' should be cotenable with '*P*' is needed if counterfactuals are not to be trivialized. Without it, any counterfactual at all would be automatically assertable. For '*P* and not-*P*' entails any '*Q*' whatever, and hence anyone who believes '*P*' to be false would be prepared to say that there is a true statement, namely 'not-*P*', whose conjunction with '*P*' entails '*Q*', whatever '*Q*' may be, and hence, in terms of this analysis without the qualification, that if *P* had been the case, *Q* would have been. But in fact we use counterfactuals discriminatingly, we accept some and not others; so we must mean more by a counterfactual than what would hold automatically, merely because the antecedent was false. The cotenability qualification is intended to provide for this by ensuring that 'not-*P*' for a start, and other trivializing conjuncts as well, are not suitable candidates for the role of '*S*' in our formula. But though we can see, at least in part, what this qualification should rule out, it is not so easy to give any exact statement of when '*S*' is cotenable with '*P*' or any criterion of cotenability that will not be viciously circular.

A closely-related account treats conditionals as condensed or telescoped arguments. Consider 'If you swim there, you will be swept out to sea', and compare it with the following argument. 'Suppose that you swim there; there is a strong undertow; you are an indifferent swimmer; there are no lifesavers or rescue apparatus around; so you will be swept out to sea.' The if-sentence is an abbreviation of the argument, leaving out the auxiliary premisses. Certainly the argument gives the sort of ground on which the conditional would be advanced; but we can understand the conditional without knowing its grounds. It is therefore like the incomplete argument: 'Suppose that you swim there; then (in view of certain unspecified facts) you will be swept out to sea.'

This account is different from the fourth, metalinguistic, one. It equates the conditional not with a higher level statement about an argument, but with the (telescoped) argument itself. To advance the conditional is to run through a condensation of the argument. We can do this without specifying, even without being able to specify, the other premisses and intermediate steps; we may jump from the supposition to the conclusion with the help of some unformulated beliefs. Similarly, we can understand a conditional used by someone else as a telescoped argument without being able to complete either his argument or one of our own. And of course the argument is not positively argued, but rather entertained; the key premiss, the antecedent of the conditional, being only supposed and not asserted.

It is a surprising and perhaps objectionable result of this interpretation that conditionals, being arguments, cannot be strictly true or false. But a compensating advantage is that the problem of cotenability disappears. Someone who uses a conditional may indeed be relying on other premisses; but if so he is not *saying that they are cotenable* with the supposition, but is merely *in fact retaining them* for use along with the supposition.

A sixth account is developed in terms of possible worlds. To say 'If *P*, *Q*' is to say that of the possible worlds in which '*P*' holds, all those closest to the actual world are ones in which '*Q*' holds too. This view brings out and builds upon what does indeed seem to be a correct and fundamental point about the word 'if', that it is used in contemplating *possibilities*, states of affairs which we do not know to be realized, or about which we do not know whether they will be realized, or (with counterfactuals) states of affairs which we believe either not to have been or not to be realized. Also, provided that we can get over the initial hurdle of setting up a system of possible worlds and relations of closeness between them and the actual world, this approach gives an attractively definite meaning to if-sentences and a systematic way of deciding disputed questions about their logic and their relations to other, e.g. modal, statements. It also provides a systematic relationship between different sorts of conditional. A strict conditional '$P \prec Q$' holds when '$P \supset Q$' is true in all logically possible worlds; an open conditional or a counterfactual when there is a 'sphere of accessibility or closeness', *S*, round the actual world, such that '*P*' is true in some possible worlds in *S* and '$P \supset Q$' is true in all possible worlds in *S*, the counterfactual being appropriate only if *P* is not true in the actual world; and the material conditional $P \supset Q$ holds, of course, provided only that it is true in the actual world.

Although these six accounts are rivals, several of them are closely related to one another, and it would be possible to combine two or more of them in various ways. It might well turn out that no interpretation will account for all if-sentences and conditionals, and that some should be interpreted in one way, some in another. But before we discuss these interpretations further, it is worth while to survey as widely as possible the different kinds of conditionals and if-sentences and some related forms, to see what range of cases an adequate account would have to cover.

§ 2. *Kinds of Conditionals and If-sentences*

First, and centrally, there are open future conditionals. Some of these are scientific and law-governed: 'If a plane mirror is placed here, that ray of light will be reflected in this direction.'

Some, though not law-governed, are still rational: 'If that party gains power, the economy will soon be in a mess.' But even ones which are not rational may be seriously put forward and understood: 'If you open that door a dragon will leap out', 'If you touch that something dreadful will happen.' There are also open past and present conditionals: 'If the Norsemen did settle in America, they didn't stay there long'; 'If he's out in this storm, he'll be soaked.'

Then there are general conditionals: 'If a politician survives, he becomes an elder statesman.'

Again, there are what Latin grammars used to call possible but improbable conditionals: 'If Israel and the Arab states were to make a genuine peace, the canal would be reopened.' The term 'subjunctive conditionals' might well be kept for these, but it is often used rather, or as well, for those that I have been calling counterfactuals. Neither term is particularly happy. 'If such-and-such had happened . . .' or 'Had it happened . . .' are not very obviously subjunctive in English, though 'If such-and-such were happening now . . .' is. And 'If this were to happen . . .' is also obviously subjunctive, and yet it belongs to a different class from the counterfactuals. But equally it is misleading to say that the conditional as a whole is contrary to fact, or even that the antecedent and perhaps the consequent separately are so. It is not their falsity in fact that puts a 'counterfactual' conditional into this special class, but the user's expressing, in the form of words he uses, his belief that the antecedent is false. 'If he had lied to me even once I should never have believed him again' remains a counterfactual even if it turns out that he told quite a few lies, whereas 'If he's out in this storm, he'll be soaked' remains an open conditional even if the man referred to has stayed indoors and warm and dry.

It is doubtful whether material conditionals occur in ordinary speech, though we have still to examine the thesis that ordinary conditionals are material ones after all. But, of course, 'If . . . then . . .' is used by logicians as a way of reading '. . . $\supset$ . . .' in words. And there are idiomatic expressions which seem to require a 'material' interpretation at any rate in preference to any consequentialist, metalinguistic, possible worlds, or telescoped argument account. 'If that horse wins I'll eat my hat' might be taken literally as a conditional offer or undertaking, though the point of making such a fantastic offer is just to indicate the speaker's certainty that he will not be called on to fulfil it. But 'If that's a Ming vase I'm a Dutchman' seems to be asserted merely because the speaker is sure that it's not a Ming vase, and the only connection that he is claiming between the antecedent and the consequent is that both are false. Even this, however, could be taken as a way of using the logical powers of the conditional form: 'If *P*, *Q*' in conjunction with 'not-*Q*' entails 'not-*P*', and so where 'not-*Q*' is a known and obvious truth 'If *P*, *Q*' can serve as an indirect but dramatic way of saying 'not-*P*'. Similarly, 'If wishes were horses, beggars would ride' serves as a way of condemning vain and idle wishes. A parallel construction is found in 'If I've told you that once, I've told you fifty times', where the obvious truth that I have told you once enables the conditional to serve as an indirect but strong assertion that I have told you fifty times. This uses the *modus ponens* power of 'If *P*, *Q*' in the same way that the Dutchman and beggars examples use its *modus tollens* power.

As J. L. Austin pointed out,[4] there are sentences in which the if-clause seems redundant, where '*Q*, if *P*' seems to entail '*Q*' outright: 'I can if I choose' seems to entail 'I can', and 'There are biscuits on the sideboard if you want them' entails that there are, simply and absolutely, biscuits on the sideboard.

'Even if Boycott goes cheaply, England will win'; 'I wouldn't marry him if he were the last man on earth'; 'He couldn't stop making money if he tried'; it may be argued that these are not conditionals, for the main clause is clearly not made conditional upon the if-clause. Still, they are undoubtedly if-sentences, and although the phrase 'even if' may be used, the second and third examples show that it need not be, that in appropriate contexts the work can be done by 'if' on its own. A comprehensive account of if-sentences should be able to deal with these, and to relate them intelligibly to the others. Like the examples in the last paragraph, these '*Q*, (even) if *P*' forms seem to entail '*Q*' itself; and they can be classified as open, counterfactual, and possible-but-improbable in the same way as ordinary conditionals. There are even cases where the 'even if' clause strongly suggests that the antecedent is fulfilled, so that 'even if' becomes equivalent to 'although': 'Even if she is fat, she's still pretty.'

[4] 'Ifs and Cans', *Philosophical Papers*, pp. 153–80.

And 'if' by itself also has this concessive use: 'He is sound, if unimaginative.'

Other uses of 'if', which go even further outside the range of what we would call conditional uses, include the optative: 'If only I had ten thousand pounds!' That it is not merely an accident that 'if' is so used seems to be borne out by the fact that εἴθε and εἰ γάρ do the same job in Greek.

So far we have concentrated on conditional statements; but there are also conditional commands—'If anything moves, shoot'—and conditional questions—'If you go abroad this summer, where will you go?' The conditionalizing of an awkward question can also be used to suggest a strong denial of the antecedent: 'If Taffy didn't take the money, where did it go?' This is a bit like the Dutchman idiom, but not quite like it, since in this case there *is* a connection between the antecedent and the consequent: Taffy's having taken the money would explain its disappearance, but if this possibility is excluded the problem of explaining its disappearance becomes acute.

We may note also the use of 'if' as equivalent to 'whether' in indirect questions: 'He asked me if I'd met anyone on the way'; 'You must remember if you paid it.'

If-sentences of various kinds can also, subject to some limitations, be embedded in other contexts. 'I'm sure that if you swim there you'll be swept out to sea'; 'He thinks that if Kennedy had lived he would have made a worse mess of things than Johnson did'; 'He told me to shoot if anything moved'; 'He asked where the money went if Taffy didn't take it'; 'It's true that if he had told me I wouldn't have believed him.' 'It's not the case that a dragon will leap out if you open that door' is stilted, but usable. One if-clause can be inserted into the consequent of another: 'If you insult him, then, if he has any spirit at all, he'll hit you'; but hardly, I think, into the antecedent of another: such constructions as 'If if you take the rook you'll be checkmated then you might as well resign' must be very rare outside textbooks of formal logic.

Counterfactuals and 'improbable' conditionals can be expressed without using an 'if' at all: 'Had he come, he would have enjoyed himself'; 'Were Israel and the Arabs to make peace, the canal would be reopened.' 'He was there all the time, had I but known' is a counterfactual of the Austinian or biscuits-on-the-sideboard class. Open conditionals too can be expressed without 'if': 'Laugh and the world laughs with you', 'One step closer, and I shoot!', 'Touch it, and it stings', 'Heads I win', 'In winter, I go round to the grocer and ask for an old box; I get one, we have a fire; I don't get one, we sit in our overcoats.'

Sentences constructed with 'since' rather than 'if' are not strictly conditionals; but they are related so closely to conditionals that there is some plausibility in Goodman's proposal to call them 'factual conditionals', by contrast with counterfactual ones. Comparing 'If he came he enjoyed himself', 'If he had come he would have enjoyed himself', and 'Since he came he enjoyed himself', it is plausible to say that while the second differs from the first only by adding the suggestion that he did not come, the third differs from the first only by adding the suggestion that he did come, and hence, in view of the first, that he enjoyed himself.

How well can the six rival interpretations outlined above cope with this range of uses? It would be pleasing to find a single account that would cover all current uses of 'if' and perhaps the related forms as well, but since this may not be possible we should note how much of the ground each interpretation could cover, if it could not cover it all.

§ 3. *The Materialist Account*

Can ordinary if-sentences be analysed as material conditionals? It is one question whether the corresponding material conditional can be correctly or appropriately used whenever and only when an if-sentence is correctly or appropriately used; it is another question whether to use an if-sentence is in effect to use the corresponding material conditional. We should take these questions separately. First, there is no doubt that whenever any sort of if-sentence is correctly or appropriately used assertively (this excludes conditional questions and commands and embedded if-constructions), the user must be prepared to assert the material conditional. The only thing that would make him unwilling to do this would be the belief that the antecedent was true and the consequent false, or at least not being sufficiently sure that this conjunction did not hold; and this would make it wrong for him to assert the if-statement. You cannot properly say, seriously, that if *P*, *Q*, or that even if *P*, *Q*, in any special way at all, if you think that it may be that '*P*' is true and '*Q*' false. The appropriateness of the material conditional is at least a *necessary* condition for the appropriateness of the if-sentence. But it seems not to be a *sufficient* condition for this. If we believe that the material conditional '$P \supset Q$' holds, say because we believe that '*P*' is false or because we believe that '*Q*' is true, but we also believe that there is no connection whatever between '*P*' and '*Q*' or what they describe, and we are not using the special Dutchman sort of idiom or its 'If I've told you once' counterpart, then we shall not find it appropriate to say 'If *P*, *Q*' except when we are doing formal logic and reading '. . . ⊃ . . .' as 'If . . . , . . .'. '$(2+2 = 5) \supset$ (It's raining)' is in order, but 'If $2+2 = 5$ then it's raining' is not.

However, a case can still be made out in defence of the materialist interpretation. It has been developed particularly by H. P. Grice[5] as part of a general theory which centres on the distinction between what is said and what is merely 'implicated' (that is, implied, suggested, etc.) between conventional meaning and (in particular) 'conversational implicature'. The need for such a distinction is illustrated by mistakes that linguistic philosophers were very prone to make about twenty years ago—e.g. Ryle's view that 'voluntary' and 'involuntary' are used (mainly) of wrong actions,[6] and that it is absurd to call correct or admirable performances either voluntary or involuntary, Wittgenstein's views that only something other than a knife and fork can be seen as a knife and fork,[7] and that we can't try to do things that are easy and unspectacular,[8] and so on. These are errors, and the general form of the error is this: certain conditions which must be fulfilled if an utterance is to have the sort of point it normally has are wrongly taken either as part of the meaning of the utterance or as prerequisites for its having a truth-value. There are, Grice says, certain general conversational principles, the basic one being a Cooperation Principle with such corollaries as 'Make your contribution as informative

[5] In his William James Lectures at Harvard University (unpublished).
[6] *The Concept of Mind*, pp. 69.
[7] *Philosophical Investigations*, II, xi.
[8] Op. cit., I, 622–3.

as is required (for the current purpose of the exchange)' and 'Be relevant'.

More important, for Grice, than simple obedience to these principles is the way in which we can work out 'conversational implicatures' which are non-trivially necessary in order that these principles should be obeyed. For example, if a reviewer says 'Miss *X* produced a series of sounds which corresponded closely with the score of "Home Sweet Home" ', instead of saying 'Miss *X* sang "Home Sweet Home" ', he has apparently violated the maxim, 'Be brief or succinct'. If he hasn't violated it, he must be indicating some striking difference between Miss *X*'s performance and one to which the simpler description would be applied. So it is a conversational implicature of the reviewer's remark that there was some hideous defect in the performance.

Grice uses this approach to criticize the view that the meaning of 'If *P*, *Q*' differs from that of '$P \supset Q$' by adding that *P* would, in the circumstances, be a good reason for *Q*, or that there are non-truth-functional grounds for accepting '$P \supset Q$'. Such additions are to be explained away as conversational implicatures of 'If *P*, *Q*', not as part of its conventional meaning, which is the same as that of '$P \supset Q$'. This is explained in two ways. The simpler is that the principle of conversational helpfulness would be violated if a speaker knew that *Q*, or that not-*P*, and yet said only something equivalent to 'Either not-*P* or *Q*'. So if someone does say 'If *P*, *Q*', which on Grice's materialist theory is equivalent to '$P \supset Q$' and hence to 'Either not-*P* or *Q*', it is a conversational implicature of his utterance that he does not know that *Q*, or that not-*P*, and hence that his assertion that either not-*P* or *Q* must have non-truth-functional grounds, that *P* must somehow be a good reason for *Q*. We can put this in other words. One does not make a weaker statement where one could make a stronger, unless one is being deliberately evasive or unhelpful—e.g. teasing, under hostile interrogation, or, while teaching, trying to make the pupil do some of the work. Without some special reason we would not say 'Jones is either in Edinburgh or in Tokyo' if we knew he was in Edinburgh, and the same principle ensures that without some special reason we would not say 'If *P*, *Q*' if we already believed that not-*P* or believed that *Q*, even on the assumption that 'If *P*, *Q*' means the same as '$P \supset Q$'. So the use of 'If *P*, *Q*' would normally carry the consequentialist interpretation as an implicature: we can then explain the evidence to which the consequentialist appeals while relying on a materialist view of the meaning of 'If *P*, *Q*'. It will be natural to use 'If *P*, *Q*' only where '$P \supset Q$' is the strongest statement we can make, and this will be precisely where we think that there is *some* connection between them which rules out the conjunction of *P* with not-*Q*.

Grice's second, deeper, explanation is meant to explain why we have the special form 'If *P*, *Q*', if it is merely equivalent to 'Either not-*P* or *Q*'. He suggests that this form is developed to facilitate the inference from this premiss, along with a second premiss '*P*', to '*Q*'. This inference is valid with the first premiss 'Either not-*P* or *Q*', but it is convenient to have an equivalent formulation in which the second premiss, '*P*', already appears un-negated in the first, and this is just what the form 'If *P*, *Q*' provides. The intended employment of the form 'If *P*, *Q*' is in such an inference, so that form will be inappropriate—though what it says will not be false—in circumstances when no such inference would be in order, either because we already know that *Q* or because we know that not-*P*.

This is an ingenious and forceful defence. It must be conceded that the expressions which a materialist equates with 'If *P*, *Q*', such as 'Either not-*P* or *Q*' carry just this conversational implicature of consequentiality: e.g. 'Either you move your car or I call a policeman.' If 'If *P*, *Q*' did have the basic meaning $P \supset Q$, it would be standardly used with just that suggestion of some kind of connection between *P* and *Q* to which the consequentialist account draws attention. Grice's argument is directed primarily against a consequentialist theory of if-sentences, and against this it is extremely forceful: it undermines the consequentialist case by explaining away the main evidence on which it rests. As an explanation of the use of open and general conditionals it would be hard to fault it.

But it is less happy with other if-sentences. It has great difficulty in accommodating counterfactuals. Whenever someone says 'If it had been that *P*, it would have been that *Q*', he is prepared to assert that not-*P*, and hence he must be prepared to assert that $P \supset R$, whatever '*R*' may be. Yet we use counterfactuals discriminatingly: someone may say 'Had it been that

P it would have been that *Q*' but not (with the very same '*P*') 'Had it been that *P* it would have been that *R*'. How is this to be explained on the assumption that the conditional part of a counterfactual (that is, everything in it besides the suggestion that the antecedent is unfulfilled) is no more than a material conditional? An extension of the Gricean theory can indeed be devised. Since the second conjunct in 'Not-*P* and if *P*, *Q*' would, on this assumption, be redundant, there will be a conversational implicature of the use of this (or of anything with the same meaning) that there is some special point in saying 'If *P*, *Q*' over and above its being made true by not-*P*, and hence that there is some connection between *P* and *Q* which would rule out the conjunction of *P* with not-*Q* even if that were not ruled out simply by the falsity of '*P*'. An analogous thesis would be that if anyone says 'The baby has a head', since he could hardly call it a baby unless it had one head, his utterance will carry the conversational implicature that the baby has two heads.

This is not impossible, but it is so fantastic as to be implausible; it would be much better if we could account in a less roundabout way for counterfactuals having the sort of force that they clearly have and for our using them with the kind of discrimination that we do display. Also, the conversational implicature in this case calls for expression by an even-if sentence which may require a non-materialist interpretation.

In fact, even-if sentences in general constitute another difficulty for the Gricean account. Someone who asserts such an open even-if sentence as 'Even if Boycott goes cheaply England will win' is prepared to assert the consequent, 'England will win', on its own, so he is indeed prepared to assert '(Boycott goes cheaply) ⊃ (England will win)'. But the converse does not hold. Someone who is convinced that Boycott will score a century but that England will lose is prepared to assert the material conditional just mentioned, but he is clearly not prepared to assert 'Even if Boycott goes cheaply, England will win'. The latter sentence must have some other meaning or force, but the implicature that England's winning is some sort of consequence of Boycott's going cheaply is not now appropriate. Even-if sentences would be an exception to a consequentialist account too.

The materialist could suggest that 'Even if *P*, *Q*' means $(-P \supset Q)\&(P \supset Q)$. Anyone who asserts the former would also be prepared to assert the latter, and anyone who both asserts the latter and is willing to suggest that *Q* goes more easily or more obviously with not-*P* than with *P* would be prepared to assert the former. Can we explain how 'Even if *P*, *Q*' comes to have this force on the assumption that 'If *P*, *Q*' basically means $P \supset Q$, that is, that it is equivalent to 'Not (*P* and not-*Q*)'? Perhaps we can. Reading 'even' as 'and equally', 'Even if *P*, *Q*' would mean 'And equally not (*P* and not-*Q*)'. Used in a context where it was natural to assume that not (not-*P* and not-*Q*)—e.g. that not (Boycott makes a good score and England fails to win)—the 'and equally' may well suggest that it is to this that 'not (*P* and not-*Q*)' is to be added, so that 'Even if *P*, *Q*' comes to have the force of 'Not (not-*P* and not-*Q*) and not (*P* and not-*Q*)', the second conjunct being asserted, the first merely suggested. This is all a bit forced, but it is possible.

When 'if' introduces conditional commands and questions, these are often most naturally construed so that the antecedent is in the indicative mood, and only the consequent is imperative or interrogative. (There are some cases where the antecedent is within the command or question, which can be construed as 'Bring it about that if *P*, *Q*' or as 'Is it the case that if *P*, *Q*?' but there are many for which this sort of meaning is inappropriate—e.g. 'If it rains, stay indoors' and 'If he comes, what will he do?') In these constructions, 'if' cannot be doing the job done by '$\supset$', and there seems to be no way of explaining the job that it does as an implicature derived from the materialist meaning.

The outcome, then, is that while Grice's approach copes easily with the apparent consequential force of open and general conditionals, it has less plausibility as an account of counterfactuals; a similar defence of a materialist interpretation of even-if sentences is possible but a bit forced, and conditional commands and questions seem not to be covered by the materialist view.

In any case, there is a radical difference between the anti-Rylean, anti-Wittgensteinian examples where Grice's distinction between conventional meaning and conversational implicature seems clearly applicable and the corresponding treatment of if-sentences. The predicates '. . . is voluntary',

'. . . is involuntary', '. . . is seen as . . .' and so on have obvious prima facie meanings, to which their grammatical structure and etymology point very plainly. E.g., to say that an action (under a certain description) was voluntary is to say that the agent's will (*voluntas*) was causally relevant: either the agent would not have done what he did if he had not wanted the result indicated by the chosen description, or at least he would not have done it if he had wanted strongly enough that this result should not come about. With this meaning, '. . . is voluntary' *would* apply, but uninterestingly, to cases to which 1950-ish philosophy would not allow it to apply. The restriction on which Ryle insists is eminently explainable by the conversational principle that one does not say what is obvious and relevant to nothing that is in hand. But there is no such prima facie etymological or grammatical reason for supposing that 'if' means $\supset$, and that apparent divergences from this use call for explanation.

Also, there is some uncertainty about the exact character of Grice's thesis. Is it a psychological hypothesis or only an 'as if' account? It is plausible to suggest that when one reads the reviewer's remark about Miss *X*'s singing one notices the oddity, supplies an explanation, and so gets the message. But a serious psychological hypothesis of the same sort about 'If *P*, *Q*' would require that a hearer should register the $P \supset Q$ meaning, realize that the remark would violate the principle of helpfulness if the speaker knew that not-*P* or knew that *Q*, supply an explanation of the speaker's using the apparently unhelpful form, and so get the consequentialist suggestion. But it is implausible to claim that English hearers generally do recognize 'if' as meaning primarily $\supset$. The use is quite different with 'Either you move your car or I call a policeman'; it is quite plausible to assume that the ordinary hearer first registers 'or' as equivalent to the logician's '$\vee$'. If, on the other hand, Grice is merely giving an hypothesis of the 'as if' variety, saying that one can give a neat exposition of the uses of 'if' by starting from the material conditional, but not claiming that this is how 'if' comes (either historically or psychologically) to have these uses, then the only relevant criticism will be to point to cases, like the conditional commands and questions, for which this does not provide a neat account.

I shall argue (in § 9) that an account based on the notion that 'if' means suppose that . . . is both prima facie more plausible and covers the ground more adequately: it can explain the consequential force that many conditionals seem to have not indeed as a conversational implicature but as something similar: the most obvious, though not the only possible, point of supposing that *P* and then asserting that *Q* is to argue, or to hint that one could argue, from *P* to *Q*.

It might seem to be a merit of the materialist account that it provides all if-sentences with truth-values. But it does so in such a crude way (e.g. all counterfactuals being true if the speaker is right in assuming the antecedent to be unfulfilled) that it would still have to be admitted that much of the point of our use of counterfactuals escapes this method of determining truth-value, and is perhaps not susceptible of truth and falsehood at all.

§ 4. *The Logical Powers Account*

Can we say that the meaning of conditional sentences is given by their logical powers? Certainly they *do* have logical powers, and their meaning must be such as to confer these powers on them, but it is another thing to say that their meaning can be completely explained in terms of these powers. The latter is, indeed, ambiguous. As we noted, it might mean that the if-sentence is just a something-or-other that has the right entailments, or it might mean that the if-sentence is itself just an inference-licence, e.g. that to say 'If *P*, *Q*' is just to authorize someone to infer '*Q*' from '*P*'. But the first alternative seems incoherent; it must collapse either into the claim that the if-sentence has some further as yet unexplained meaning which confers these logical powers, or into the second alternative; it is therefore the latter that we need to examine now.

We can agree that anyone who believes that if *P*, *Q* will be ready, on learning that *P*, to infer that *Q*. But it is difficult to describe his state of mind without using a conditional: he is disposed to infer that *Q if* he learns that *P*. Equally, we cannot understand *giving* an inference-licence except in terms of conditionals: to give such a licence is to say 'If you learn that *P*, you may infer that *Q*'. As an analysis of conditionals, the inference-licence account seems viciously circular. It is, indeed, plausible to say that the Dutchman type of idiom exploits the logical powers of 'if', especially the *modus tollens* one: it is like an enthymeme, inviting the completion 'The speaker is not a Dutchman; therefore not . . .' (whatever the antecedent may have been). But it does not follow that the sentence should be construed as saying just 'From . . . you may infer that I am a Dutchman'. In any case, this analysis, like the material conditional one, fails for counterfactuals. Anyone who says 'Had it been that *P* it would have been that *Q*' has no use for the inference '*P*, therefore *Q*', since he cannot coherently assert *P*. Equally he has no use for the inference 'not-*Q*, therefore not-*P*', since although he believes 'not-*Q*' he also already believes 'not-*P*' independently. So neither of the most obvious and basic logical powers of 'if' is relevant here. We must not overstate this point. It is not that the counterfactual does not have the usual logical powers. 'Had it been that *P* it would have been that *Q*' together with '*P*' does entail '*Q*'—and not merely because 'not-*P*' together with '*P*' entails *Q*, but because of the conditional component in the counterfactual. And similarly this counterfactual together with 'not-*Q*' entails 'not-*P*', and not merely because the counterfactual itself suggests that not-*P*. But although these powers are there, they are redundant. They cannot be used to give us anything we do not already have. So the point of using the counterfactual cannot be to give either of these inference-licences. Again, 'if' cannot be giving an inference-licence in the biscuits type of case or in 'I can if I choose'. Nor could exactly this be done by conditional questions or commands, though it might be said that something analogous is done by these: it is when and only when the antecedent is satisfied that the command or question in the consequent comes into force.

§ 5. *The Consequentialist Account*

We come next to the consequentialist type of interpretation. One variant of this is the view that the 'if' in any sentence represents some specific kind of consequence—that there are logical 'ifs', causal 'ifs', and presumably moral, legal, religious, and magical 'ifs', as well as an 'if' of the kind of artificial consequence set up in our example of the quiz contest. But this is utterly implausible. It would be almost as absurd to suggest that what conjoins two legal provisions must be a legal 'and'. There may be reasons of many different sorts for saying something of the form 'If *P*, *Q*', but there is no need for the reasons to infect the if-sentence they support.

A more plausible variant of this account is that the 'if' in any sentence merely says that there is some consequential relation between the items it connects. 'If *P*, *Q*' would be equivalent to '*Q* is some sort of consequence of *P*', or '*P* somehow would ensure *Q*'. And this equivalence seems to hold in all the central cases. Anyone who says 'If *P*, *Q*' is normally prepared to say that *P* somehow would ensure *Q*, and anyone who is prepared to say that *P* would ensure *Q* will be prepared to say that if *P*, *Q*. In particular, this account covers counterfactuals, which the materialist and the logical powers interpretations did not cover. But this view also faces serious difficulties. One is the danger of circularity. Can we explain what it is for *P* to *ensure* *Q*, or for *Q* to be a *consequence* of *P*, without using the very conditionals that we were hoping to analyse? Perhaps we could explain the general concept of a consequence by giving examples of logical consequence, causal consequence, and so on. But will not conditionals crop up in the explanation of the specific types of consequence? In particular, one of the main reasons for seeking an adequate account of conditionals, especially counterfactuals, is that they seem to enter into the analysis of causal statements. And even if we could break out of the circle, the consequentialist account faces another radical objection. It seems not to cover a number of the standard, though relatively peripheral, uses of 'if'. 'I can if I choose' does not say that my being able would be a consequence of my choosing. But we might argue that this formula is elliptical, that it should be construed as 'I am able to (do *X* if I choose)', i.e. I am able to make my doing *X* a consequence of my choosing. Similarly, there being biscuits on the sideboard is not ensured by your wanting them; but perhaps this too is elliptical for 'There are biscuits on the sideboard; take some if you want some'. This leads to the question whether the consequentialist account can cover conditional commands—and, analogously, conditional questions. We should presumably have to take these as saying something like '*P* has as a consequence the being in force of the following command/question: ". . ." ' And now we are being forced into a metalinguistic account. Whereas we can say '*P* ensures that *Q*' while talking only about the events, states of affairs, or what not, *P* and *Q*, without moving into a metalanguage and talking about the statements or sentences '*P*' and '*Q*', this does not work for conditional commands or questions. '*P* ensures that bring it about that *Q*' and '*P* ensures that is it that *Q*' do not make sense. What is ensured, what can be a consequence, cannot be the actual content of a command or a question, it can only be the *giving* of the command or the *asking* of the question. Again, 'Even if *P*, *Q*' does not say that *P* would ensure that *Q* or that *Q* is a consequence of *P*. Someone who says 'Even if Boycott goes cheaply England will win' is not suggesting that Boycott's going cheaply will ensure England's winning, or that the latter is any sort of consequence of the former.

The Dutchman idiom is commonly taken as an exception to the consequentialist interpretation: but a case can be made out on the other side. It is most natural to say 'If *P* then I'm a Dutchman' where '*P*' would in some way falsify the speaker's judgement about horse-racing, antiques, and so on, and 'Dutchman' symbolizes a stupid person, one lacking in judgement. So my being a metaphorical Dutchman is after all a kind of consequence of *P*. Similarly, it could be argued that eating one's hat is a metaphorical description of a state of extreme surprise, so that 'If *P* then I'll eat my hat' is only picturesque for 'If *P* then I'll be surprised', which could be consequential. But I am not sure that this account explains the idiom completely, and it seems not to cover the counterpart 'If I've told you once I've told you fifty times'.

In short, there are several peripheral cases in which if-sentences are used but it seems that no ground-consequent relation is in question; and even in the many central uses of 'if' where such a relation is involved, it does not seem possible without circularity to analyse the if-sentence in terms of this relation.

§ 6. *The Metalinguistic Account*

We saw that the consequentialist account had to move into the metalanguage in dealing with conditional questions and commands. Is a generally metalinguistic account acceptable? We noted, earlier, that there was a problem about deciding when other premisses are cotenable with the antecedent. But perhaps this is a spurious difficulty. For if we say that 'If *P*, *Q*' is equivalent to 'Either "*P*" entails "*Q*" or there is some statement or set of statements "*S*" which is true, and cotenable with "*P*", and the conjunction of which with "*P*" entails "*Q*" ', it is the *speaker*, the user of the if-sentence, who is claiming the cotenability of '*S*' with '*P*', not the *analyst*. The analyst, *qua* analyst, is merely claiming that the users of if-sentences implicitly use a concept of cotenability. He is committed only to a verbal explanation of this term, which he can give thus: to say that '*S*' is cotenable with '*P*' is to say that it is reasonable to combine the belief that *S* with the supposition that *P*. The analyst need not himself give any firm criteria for cotenability, or himself assert the distinctive rationality of some combinations of beliefs with suppositions. On the other hand, if the analyst takes a positively sceptical view about this, and admits that no satisfactory criteria for cotenability can be given, he is then committed to the view that if-statements in general are somewhat arbitrary, that the use of them cannot be rationally defended. And this might be an alarming sort of scepticism. After all, we are constantly using if-sentences; I have myself used far more in this chapter than I have mentioned or quoted as examples.

The metalinguistic account has the advantage over the consequentialist one that it can accommodate 'even if'. In saying 'Even if Boycott goes cheaply, England will win' one might well be saying that there is a set of true statements '*S*', cotenable with 'Boycott goes cheaply', such that the conjunction of these entails 'England will win'. This conclusion will no doubt be entailed by '*S*' alone; 'Boycott goes cheaply' will not be helping to generate this conclusion; the 'even if' merely indicates that it will not hinder it either, that the '*S*' in question is strong enough to produce the conclusion even when encumbered by this hostile supposition.

Apart from this, the metalinguistic account covers nearly all the cases that the consequentialist one did, including possibly the Dutchman idiom and the biscuits examples, and it can also cope with conditional questions and commands. I say 'nearly all', because it fails with irrational if-sentences. Someone may say 'If you touch that something dreadful will happen' and believe it, and be afraid to touch 'that' for this reason, and yet not be prepared to claim that there is a set of true statements '*S*' cotenable with 'You touch that' which, with the latter, entails that something dreadful will happen. One may seriously put forward an if-statement without claiming the sort of rational backing that the metalinguistic analysis would involve. Indeed, I think that primitive, unsophisticated causal statements and beliefs involve counterfactuals which escape the metalinguistic analysis in this way. A child or a savage who notices a striking intrusion into an otherwise peaceful and relatively stable state of affairs, followed shortly by another striking occurrence, may well think that if the intrusion had not occurred, the second occurrence would not have happened either. But he is not *inferring* this conclusion from his supposition along with some other set of statements, let alone saying that the conclusion is *entailed* by such a conjunction. His thinking is analogical and imaginative: it is simply by projecting the previous peaceful state of affairs forward in time that he gets his picture of how it would now be if the intrusion had not occurred.

The radical objection to the metalinguistic account is that although it could cover much of the ground, it simply does not ring true. Using an if-sentence is a ground-floor linguistic performance, not a higher level one. To say 'If that party gains power the economy will be in a mess' is to talk about the (possible) political and economic events themselves, not about what can be said about them. And though if-sentences are not the simplest parts of ordinary language, they are not very elaborate parts either. It seems clear that someone who does not or even cannot reach the degree of sophistication required for talk about talk may well use if-sentences. As Hume rightly insisted, causal concepts (which require conditionals for their elucidation) are not the preserve of scientists and philosophers, but are shared by children and ignorant peasants, and perhaps by some animals as well.

§ 7. *The Condensed Argument Account*

The telescoped argument interpretation takes account of this last objection to the metalinguistic view. It has, as we saw, the advantage of not requiring any firm criteria for cotenability of assumptions, but we have also seen that the metalinguistic view can claim this advantage too. But either view can avoid the problem of cotenability only by abandoning the claim that conditionals are true in a strict sense. The telescoped argument view, while it is more natural in that it interprets conditionals as ground-floor discourse, is less natural in that it says that they are not statements at all. On the face of it, an if-sentence appears to make a statement of some sort rather than to formulate an argument.

The telescoped argument view also shares both some of the merits and some of the limitations of the metalinguistic one. It readily accounts for most open conditionals and counterfactuals. With its weaker version of cotenability, that is, by taking note of what existing beliefs it is reasonable to retain along with a belief-contravening supposition, it can, as I have argued elsewhere, explain our rejection of implausible counterfactuals and resolve the problem of competing counterfactuals. Its greatest merit is that it can explain, in a totally non-mysterious way, why causal laws sustain counterfactuals whereas generalizations of fact do not. But all these points will be considered again later. The argument view can equally well accommodate even-if sentences. In using one of these we are running through an argument which has the supposition of the antecedent as a premiss, though the conclusion is derived from the other, unstated, premisses alone, and the antecedent merely does not undermine this derivation. It has difficulty with examples of the Austinian sort. It could just cope with conditional commands if one allowed the appropriate kinds of imperative inference. For example, we could construe 'If you go out, wear your coat' somewhat as follows:

Supposed premiss: 'You go out.'
Auxiliary, unstated premisses: 'Always wear something warm when you are in cold conditions; there are cold conditions outside now; your coat is the only warm thing convenient and available for you to wear.'
Conclusion: 'Wear your coat.'

But it does not seem able to cope with conditional questions in any analogous way. And it fails with irrational conditionals like 'If you touch that something dreadful will happen', and with the counterfactuals involved in primitive causal beliefs, for the same reason that the metalinguistic view does: the thinking which these counterfactuals represent is imaginative analogizing, not inference or even condensed inference.

§ 8. *Possible Worlds*

The possible worlds theory has been developed particularly as an interpretation of counterfactuals, and as a way of supplying them with truth conditions. The most elaborate account of this kind so far put forward is that of D. K. Lewis. He posits for each world i a set of nested 'spheres of accessibility', each sphere containing various possible worlds, so that for a sphere S around i every possible world in S is more closely similar to i than is any world outside S. Given this scheme, he states the truth conditions for the counterfactual 'If it were the case that ϕ then it would be the case that ψ', which he writes '$\phi \mathrel{\Box\!\!\rightarrow} \psi$', as follows:

> '$\phi \mathrel{\Box\!\!\rightarrow} \psi$' is true at a world i if and only if either (1) no sphere around i contains any worlds in which 'ϕ' is true, or (2) some sphere S around i does contain some worlds in which 'ϕ' is true, and '$\phi \supset \psi$' is true at every world in S.

The first alternative gives a vacuously true counterfactual; it is the second that gives the sort of counterfactual that can be seriously asserted, whose antecedent can be entertained.

This account means that a counterfactual is true if in all the closest possible worlds in which the antecedent holds, the consequent holds also.

This elaborate account, presupposing a system of possible worlds and nested spheres containing them, surely does not describe what the user of a counterfactual ordinarily means or intends to convey. It must be taken as a proposed linguistic reform, as a description of something that counterfactual conditional sentences *could* mean without diverging too far from their present use, indeed while being still usable on just the same occasions on which they are now used, but which would enable them to be determinately true or false and would provide them with an exact logic and with reasons why they obey it.

We might suggest the following connection between this proposal and what counterfactuals ordinarily mean. It would be quite plausible to say that '$\phi \mathrel{\Box\!\!\rightarrow} \psi$', as ordinarily used, means 'In the possible but not actual situation where ϕ, ψ also' —I am deliberately avoiding a higher level formulation, 'where "ϕ" holds' and so on, for the reasons given in criticizing the metalinguistic view. The ordinary counterfactual does purport to describe a possible state of affairs or sequence of events. But then we realize that there is not just one possible situation where ϕ: the phrase '*the* possible . . . situation . . .' is not justified. In framing possible situations where ϕ, we cannot retain all the other features of the actual world, for various combinations of them will be logically incompatible with 'ϕ' 's holding—e.g. if the actual occurrence of not-ϕ was caused, we cannot retain both the causal law and all the causal antecedents of not-ϕ—but we can retain some so long as we abandon others, and we have a choice about what we retain. Talking, explicitly or implicitly, about 'the possible situation where ϕ' conceals this choice and prevents what we say from having a determinate truth-value. The system of possible worlds is an attempt to remove this indeterminacy, to introduce rules that make each non-vacuous counterfactual refer to a definite set of possible worlds, namely those closest to our own in some of which 'ϕ' holds.

However, the indeterminacy is not removed, but only relocated. Lewis himself admits that counterfactuals as he interprets them are like such statements as 'Seattle resembles San Francisco more closely than it resembles Los Angeles', since the truth of a counterfactual depends upon the relative closeness of possible worlds, i.e. their relative similarity to the actual world. 'If you had struck that match it would not have lit', said of a wet match, will be true if the closest possible world in which the match is struck is one where it is struck while still wet, our causal laws being still in force. But it is false if the closest possible world is one in which the striker, being knowledgeable and observant, would not strike a wet match, but takes care to dry this one before striking it. As Lewis says, the relative similarity between towns depends on whether we attach more importance to the physical surroundings, or the architecture, and so on. Likewise the relative similarity between these worlds depends on what we attach importance to. But then neither the statement about Seattle nor the counterfactual thus interpreted is capable of being true or false. The claim to have supplied truth-conditions fails. What have been supplied are still only acceptability conditions; someone who *regards* the first of the two possible worlds as more like the actual world will *accept* or *use* the counterfactual, someone who *regards* the second as more like the actual world will *reject* it. But it is hard to see that this is any advance on saying simply that someone will accept the counterfactual if he retains, along with the supposition, the present condition of the match and the ordinary causal laws, but he will reject this one, and accept the contrary counterfactual, 'If you had struck this match it would have lit', if he retains instead the prudent character of the striker.

Moreover, talk about possible worlds and spheres of accessibility cries out for further analysis. There *are* no possible worlds except the actual one: so what are we up to when we talk about them? It seems clear that this analysis would have to be given in terms of what people do in the way of considering, supposing, inserting or excluding elements in a picture, and so on. This does not mean that talk about possibilities is entirely arbitrary or fantastic. We may decide to include in our possible world unknown contingent features of the actual world—e.g. that this match should remain in whatever condition it is.[9] And if we decide to retain causal laws carried over from the actual world, this will seriously restrict the possibilities in a contingent and perhaps partly unknown way. A possible world, then, is not a dream world or fairyland in which everything is exactly and only as it is imagined to be. But it does mean that this is too shaky a foundation for the truth of counterfactual statements detached from their users.

The elaborate sort of possible worlds account fails, then, in its distinctive aim of providing truth-conditions for counterfactuals. In so far as it is defensible, it is only another way of formulating their acceptability conditions in terms of what one is prepared to retain when one introduces a belief-contravening supposition. And in any case it is a proposal for linguistic reform, not an analysis of what is ordinarily meant. The simpler thesis from which it starts, that '$\phi \;\Box\!\!\rightarrow \psi$' is equivalent to 'In the possible but not actual situation where ϕ, ψ also' is defensible as an initial analysis of the counterfactual; but this still calls for a further analysis of what is involved in assertions about

[9] Stalnaker, op. cit., pp. 111–12.

possible situations, and this will probably fall back into one of the rival interpretations of conditionals.

We have considered the possible worlds account as applied only to counterfactuals. Would it serve as even an initial analysis of conditionals or if-sentences in general? Corresponding to what I have just called the simpler thesis about a counterfactual would be the view that an open conditional 'If *P*, *Q*' is equivalent to 'In the possible situation where *P* (which may or may not be actual), *Q* also'—for example, 'In the possible situation where he's out in this storm, he's soaked'. This seems right, as far as it goes, and it brings out the fact that the conditional will not be satisfied merely by his not being out in the storm. A more elaborate account, analogous to that of Lewis for counterfactuals, would give these truth-conditions for an open conditional:

'If *P*, *Q*' is true, non-vacuously, at a world *i* if some sphere *S* round *i* contains some worlds in which '*P*' is true, and '$P \supset Q$' is true at every world in *S*.

That is, the open conditional will be non-vacuously true (1) if '*P*' holds in the actual world and '*Q*' holds also, or (2) if '*P*' does not hold in the actual world but in some other possible world, and '*Q*' holds in this possible world, while '$P \supset Q$' holds in all worlds that are at least equally close to the actual one.

This suggestion will be open to the same objections as those made against the attempt to give truth-conditions for the counterfactual, but if it is reduced to a claim to give acceptability conditions it seems correct. In particular it brings out the point that when someone says 'If *P*, *Q*' and it then turns out that not-*P*, he cannot simply say 'So I was right', but must rather say 'Well, if it had been that *P*, it would have been that *Q*'. Using an open conditional, and then learning that its antecedent is false, commits one, for the sake of consistency, to adopting the corresponding counterfactual.

None of these six accounts, then, is entirely satisfactory. Each of the first five encounters at least some if-sentences whose use it fails to accommodate or fails to explain. The sixth, possible worlds, account, in its elaborate form makes a misguided attempt at providing truth-conditions, especially for counterfactuals, and in any case must be taken as proposing a new meaning for conditionals rather than analysing their present meaning. But in its simpler form, where it equates 'If *P*, *Q*' with 'In the possible situation where *P*, *Q* also' it avoids both these objections. It clearly covers open as well as counterfactual conditionals, the latter being differentiated by the suggestion that this possible situation is not actual. It can easily explain the analogy with since-sentences, for 'Since *P*, *Q*' can be construed as 'In the possible situation where *P*, *Q* also; and *P*; hence *Q*'. The Dutchman idiom can then be construed as an enthymeme: 'In the possible situation where *P*, I'm a Dutchman' with the unstated continuation 'I'm not a Dutchman; so not-*P*'. 'If I've told you once . . .' becomes a similar enthymeme with the conclusion 'I've told you fifty times'. Even-if sentences, open or counterfactual, can be construed just as easily, and so can conditional questions and commands. So too can embedded if-sentences. What are not fully explained by this account are the concessive use of 'even if' and 'if'—'Even if she's fat, she's still pretty'; 'He is sound, if unimaginative'—also the Austinian examples, and 'if' for 'whether'. Something more should also be said about general conditionals.

This account, however, is still exposed to the criticism that it gives only an initial analysis of if-sentences, that its 'possible situations' and what is in general the contingent claim that some '*Q*' *holds in a certain possible situation* cry out for some further analysis. This analysis must be given in terms of what people do. Possible situations, or possible worlds, just because they are not actual (or may not be actual) do not stand on their own, do not exist independently. We must come back in the end to something like a speech-act analysis of statements about possible situations, and hence of the conditionals which were initially analysed in this way. People can consider possibilities; but the possibilities exist only as the contents of such considerings.

§ 9. *Supposition*

I would suggest, then, that the basic concept required for the interpretation of if-sentences is that of *supposing*. But we also need what is perhaps the obscure concept of asserting something within the scope of a supposition. This is the act-counterpart of the notions of possible worlds or possible situations. A supposition may introduce not only the single item that is supposed, but a complex picture, which is held together and partly determined by what else the supposer associates with this first item, typically by retaining and carrying over elements from the actual world. And not only fully known elements. The supposer can, as we saw with the match example, carry over *whatever* intrinsic features something actually has, including ones that he does not yet know about. He can then be committed to filling in his description of the possible situation in certain ways, to saying things within the scope of his supposition that he did not arbitrarily put there. His possible world has some autonomy, it can take on, to some degree, a life of its own; he can make discoveries and contingent assertions about it.

I want to offer, then, this general analysis: *to say* 'If *P*, *Q*' *is to assert Q within the scope of the supposition that P*. This is, of course, an analysis of the same general sort as the condensed argument view. Like it, it stresses the making of a supposition and the doing of something in relation to it. Like it, it translates 'If' as 'Suppose that . . .'. And like it, it abandons the claim that conditionals are in a strict sense statements, that they are in general any sort of descriptions that must either be fulfilled or be not fulfilled by the way things are, and hence that they are in general simply true or simply false. But my present analysis is deliberately wider than the condensed argument one. There are, as we have seen, if-sentences which do not expand into arguments with the antecedent used as a premiss, and my present analysis makes room for these. The telescoped argument account will still cover a majority of the central kinds of if-sentence, but it gives only one of several reasons that we may have for asserting something within the scope of a supposition. There are, indeed, sub-classes within the class of condensed arguments. In logically necessary if-sentences, the ones that would misleadingly be said to contain a logical 'if', the consequent will be inferred from the supposition alone. In most ordinary conditionals, including the typical counterfactuals, the consequent is inferred from the supposition in conjunction with other, unstated, premisses. In even-if-sentences, the consequent is inferred from other, unstated, premisses alone, but it is still asserted within the scope of the supposition: the 'even' indicates that though the supposition might have been expected to undermine this inference, it does not. But in other cases the consequent is not the conclusion of an inference at all. In the idiomatic 'If that's a Ming vase I'm a Dutchman' there *may*, as we have seen, be a hint of a ground-consequence relation, and so of a kind of argument from the antecedent to the consequent. But there need not be. If there is not, then the speaker is just asserting something known to be false within the scope of the supposition, and this is an indirect but rhetorically effective way of rejecting the supposition. In the parallel idiom 'If I've told you once I've told you fifty times' the speaker, by asserting the consequent within the scope of a known-to-be-true antecedent, is again indirectly, but thereby dramatically, asserting the consequent absolutely as well. The effectiveness of this device is that of an enthymeme with a suppressed conclusion: presumably a hearer is more moved by a conclusion which he has expended some effort of his own, however minimal, in reaching. In the irrational conditional, such as 'If you touch that, something dreadful will happen', the speaker is prepared to assert the consequent within the scope of the supposition, but just because he associates them; he is not able to give any argument that supports the consequent, and he may not even believe that there is any such argument to be discovered. Something similar applies, as I have said, to the counterfactuals involved in primitive, unsophisticated causal beliefs. The child may believe that if he had not touched the fire, his finger would not now be hurting, that is, he is in effect prepared to assert 'My finger is not hurting' within the scope of the known-to-be-false supposition 'I did not touch the fire', not because he can infer the consequent from the supposition plus a causal law—for he is now for the first time discovering that causal law by learning an instance that will come under it—but just by an imaginative transfer of the freedom-from-hurting from the experienced prior state (when he had not touched the fire) to the presently counterfactually imagined state of his *now* not having touched it. But with developed and sophisticated causal beliefs, held in association with a number of well-confirmed laws and regularities, a condensed argument account of the counterfactuals they involve may be in order.

The Austinian 'There are biscuits on the sideboard if you want them' can be construed in either of two ways. We might say that the speaker is prepared to assert 'There are biscuits . . .' absolutely, and *a fortiori* is prepared to assert this within the scope of the supposition; thus construed, it would be like an even-if-sentence, except that the antecedent would be seen as neutral, not as even prima facie damaging to the consequent. But it is more naturally construed thus: 'There are biscuits . . .' is asserted absolutely, outside the scope of the supposition. The supposition is introduced, but then nothing further is done within its scope. It is just left hanging. And a hanging 'Suppose that you want them . . .' naturally means almost the same as 'and perhaps you want them'. Austin called this an 'if' of doubt or hesitation. But it is one of the merits of the present account that it can explain even this as a limiting case of a suppositional 'if', not as any radically different sort of construction. Of course, as was previously suggested, this example can be assimilated to the conditional command, by being expanded into 'There are biscuits . . . take some if you want some', and hence to the more normal sort of case where something further is said within the scope of the supposition, though what is there put forward is not the apparent consequent, not what appears as the grammatical main clause. But there is no need for this expansion; the construction can be understood without it as a natural outgrowth of the standard use of 'if'.

'He was there all the time, had I but known' is to be construed similarly: an absolute assertion followed this time by a hanging counterfactual (i.e. belief-contravening) supposition. When a belief-contravening supposition is left hanging, it cannot, of course, mean 'and perhaps . . .'. The possibility that I might have known is contemplated but not as one whose fulfilment is cognitively open; and the natural reason for so contemplating it is that the speaker finds it attractive. Thus 'had I but known' comes to express a wish that I had known or a regret that I did not know, and this too is a natural outgrowth of the standard use of if-clauses and their equivalents. The optative use of 'if only', which can be either counterfactual—'If only I had told him'—or open—'If only we can keep him talking for another five minutes'—can be understood in the same way. A supposition is introduced but left hanging; and this introduction suggests that this possibility is one that the speaker likes to contemplate. An open 'if only' can thus be paraphrased, a little poetically, as 'suppose that . . . oh rapture!' But in a counterfactual 'if only' the rapture is transformed into regret by the belief or knowledge that the desirable possibility is unfulfilled.

Although 'I can if I choose' seems to resemble the biscuits and 'He was there all the time . . .' examples in that they all entail the unconditional truth of the main clause, this resemblance is superficial. They do so for very different reasons. Whereas in the other cases this is a direct outcome of the structure of the sentence—the main clause just is asserted absolutely, and is not really a consequent—'I can' *is* a consequent after all. In fact, this example, for all its notoriety, is rather a red herring in an account of if-sentences. Its peculiarities stem not at all from its containing a special sort of 'if' or even a special construction with 'if', but merely from the special meanings of 'can' and 'choose'. Basically what is said here is that on the supposition that I choose to do *X*, I am able to do *X*; this is a telescoped argument case, to be expanded by adding some such unstated premiss as 'There are no physical obstacles or limitations'. But we assume that choosing to do *X* is always within my power; this is necessarily something I can do. So, given that choosing to do *X* is, in this situation, all that is needed to make me able to do *X*—I can if I choose—and that I can choose, it follows that doing *X* also is within my power. Briefly, 'I can' follows from 'I can if I choose' because it follows from the latter in conjunction with 'I can choose', and this last is assumed to be necessarily true. But as an if-sentence this example is in no way exceptional.

We have still to explain the concessive use of 'even if' and of 'if' by itself: 'Even if she's fat, she's still pretty'; 'He is sound if unimaginative.' Whereas ordinary even-if-sentences fall clearly and easily within our schema of a supposition and something asserted within its scope, these do not. The speaker is not merely *supposing* that she is fat or that he is unimaginative, he is saying that each of them is so, but saying it obliquely, mentioning it as something that contrasts with the main assertion, but does not conflict with it, though it might have been expected to. I suggest that these are factual conditionals (borrowing Goodman's term) of the even-if group. Since-sentences are the factual conditionals corresponding to ordinary, consequential, open conditionals and counterfactuals: concessive if-sentences are those that correspond to even-if open conditionals and counterfactuals. We have already noted a possible worlds interpretation of 'Since *P*, *Q*': 'In the possible situation where *P*, *Q* also; and *P*; hence *Q*'. When we supply the further suppositional analysis for the possible-situation talk this becomes (in the consequential sub-class): 'Suppose that *P*; then (in view of . . .) *Q*; and *P*; hence *Q*.' The even-if counterpart of this would be, for example: 'Suppose that he's unimaginative; nevertheless (in view of . . .) on that supposition he's sound; he is unimaginative; but (from the above) he's sound.' To put it simply: when the speaker using an even-if construction also indicates that the antecedent is fulfilled, a concessive meaning automatically results.

The general conditional—'If a politician survives, he becomes an elder statesman'—is, of course, a way of expressing a universal proposition. In modern logic the standard way of expressing a universal proposition is to apply a quantifier to a material conditional—'$(x)(Fx \supset Gx)$'. Is the 'if' in a general conditional a material one, then? Not necessarily, since this sentence can be more naturally and directly construed as 'Take a politician at random; suppose that he survives; in that case he's an elder statesman.' We can understand why this expresses the universal proposition 'All surviving politicians become elder statesmen': the latter is precisely the additional unstated premiss needed to derive the conclusion from the supposition. To be prepared to assert of an indeterminate politician, one taken at random, on the supposition that he survives, that he becomes an elder statesman is precisely to have the corresponding universal proposition up one's sleeve.

Can we relate this general account of the meaning of 'if' to the fact that 'if' may be used, instead of 'whether', to introduce an indirect question? In an indirect *W*-question, we simply use the same interrogative pronoun, adverb, or adjective as in the corresponding direct question: 'Who is there?' turns into 'He asked who was there'. So if other direct questions—yes–no questions and stated alternative questions, such as 'Are they friendly?' 'Was she wearing shorts or a dress?'—had had an explicit interrogative operator to introduce them, something meaning 'Is-it-the-case-that', we should expect this too to be carried over into indirect speech. English has no such operator in direct speech, but 'whether' plays the corresponding role in indirect questions. I suggest that since the primary function of 'if' is to introduce a supposition, to invite us to consider a possibility, it is not too extreme a change of function for it to act as such an interrogative operator. 'He asked if they were friendly' is not too remote from 'He asked about the possibility that they were friendly'. It is worth noting that 'whether', which acts primarily as an interrogative operator in indirect yes–no and stated alternative questions, can also be used in what are undeniably double conditionals, e.g., 'He'll come whether he's invited or not', which means 'He'll come if he's invited and he'll come if he's not invited'.

This suppositional interpretation of if-sentences covers the ground, then, remarkably well. But a possible objection is that it is in effect circular and therefore trivial and unilluminating. Of course 'if' can be expanded into 'Suppose that', and the consequent can be regarded as being put forward within the scope of that supposition, but that is because to suppose that . . . is just to say 'if . . .' and to put *Q*, say, forward within the scope of the supposition that *P* is just to assert *Q* on condition that *P*, i.e. to say 'If *P*, *Q*'. To meet this criticism we must explain supposition without reducing it to saying 'if'. And we can. Philosophers in particular must be familiar with the procedure of starting off with 'Let us suppose, just for the sake of argument, that . . .' or 'Let us consider the possibility that . . .' and going on to develop the consequences of this suggestion, also bringing into the consideration contingent truths carried over from the actual world. The use of 'if' can be understood in the light of this sort of procedure; this sort of procedure does not need to be understood in terms of 'if'. In particular, to adopt a procedure of this sort is clearly not to use a material conditional or to assert a ground-consequence relation—though such relations may well be used in connection with it—which are two of the main rival accounts of 'if' to which we might have recourse if we tried to make the use of 'if' the basic thing and explain supposition in terms of it. On the other hand, this sort of procedure can be described as constructing, contemplating, or examining a possible world; but I would repeat the previous argument that talk about possibilities needs to be explained in terms of concrete human procedures, and not the other way round.

This suppositional procedure is also used, formally, in natural deduction systems of logic, where we introduce one or more assumptions and calculate in terms of them, taking care to keep some form of record, at every step, of what assumptions it relies upon and within whose scope it is asserted. This analogy may help to explain both why if-sentences can get mixed up with material conditionals and how they differ. Whenever at some stage in a formal argument we have 'Q' asserted on some set of assumptions which includes 'P', we can, by what is called conditional proof, derive from this the assertion of the material conditional '$P \supset Q$' on the remaining assumptions. A discharged assumption becomes the antecedent in a material conditional. Briefly, wherever we have the assertion of 'Q' within the scope of the supposition that 'P', we *can* assert '$P \supset Q$'. But of course this does not mean that the two things are the same: quite the reverse. It is equally true that wherever we already have '$P \supset Q$' we can introduce an assumption 'P' and then assert 'Q' within the scope of this—and, of course, also within the scope of whatever assumptions '$P \supset Q$' rested on. It would seem to follow from these points that there must be perfect coincidence between legitimate uses of suppositional if-statements and material conditionals. Whenever one is in order, so must the other be, since we can go by formally valid steps from 'Q on assumption P' to '$P \supset Q$' and vice versa. What then becomes of the arguments we used earlier to criticize the materialist interpretation of if-sentences?

We argued that someone who is convinced that Boycott will score a century but that England will lose is prepared to assert '(Boycott goes cheaply) ⊃ (England will win)' but not to assert '(Even) if Boycott goes cheaply, England will win'. But by the present argument, the former would commit him to the latter (without the word 'even') on the suppositional interpretation of 'if'. But the catch is, as so often, in the problem of what beliefs one retains along with an assumption. The speaker in question accepts '(Boycott goes cheaply) ⊃ (England will win)' just because he is sure that Boycott won't go cheaply. In these circumstances it is not reasonable for him to retain this material conditional along with the supposition that Boycott goes cheaply. Although the formal steps would allow him to assert 'England will win' on the assumption 'Boycott goes cheaply' together with his other assumption that Boycott will make a century, this is not a reasonable or coherent way of combining beliefs with suppositions. So there is after all a wedge to be driven between suppositional uses of 'if' and material conditionals, and it is with the former, not the latter, that the ordinary use of 'if' coincides. Moreover, the suppositional procedure, as we have seen, leaves room for commands and questions within the scope of a supposition, whereas these cannot fill the consequent place in a material conditional.

I would stress that this suppositional account, far from merely replacing 'if' by the near-synonym 'Suppose that', analyses the use of conditional sentences in terms of a kind of intellectual performance that is not essentially linguistic. Supposing some possibility to be realized and associating some further how-things-would-be with it is something that can be done merely by *envisaging* the possibilities and consequences in question. Among the reasons why this is important is that it makes comprehensible the otherwise puzzling fact that animals which seem not to use language—at any rate, not a language with any such grammatical structure as ours—seem capable of reasoning conditionally. A chimpanzee not strong enough to fight with others in his group for his share of some bananas was apparently capable of acting on reasoning which we should express thus: 'If I go away the others will follow me; then I shall sneak back and get some bananas before they return.'[10] This links up with the point made in § 6 in criticism of the metalinguistic account of conditionals: primitive, unsophisticated, causal beliefs involve conditionals for which a metalinguistic analysis or even a telescoped argument one would be inappropriate; the thinking expressed by conditionals in their primary employment is of a kind that can proceed without language, and certainly without a metalanguage.

There is, however, another fundamental objection to this suppositional account. It gives a 'speech-act' analysis of conditionals, in particular it explains conditionals in terms of what would probably be classified as a complex illocutionary speech act, the framing of a supposition and putting something forward within its scope. It is therefore an analysis of the same general

[10] Jane van Lawick-Goodall, *In the Shadow of Man*, p. 96. (This example was brought to my attention by Dr. J. A. Gray.)

kind as those which have explained moral language in terms of the speech acts of prescribing or commending, the meaning of 'true' in terms of re-asserting, the meaning of 'probable' in terms of expressing belief or giving guidance or partial assurance, and so on. And some philosophers have thought that this sort of analysis is wrong in principle, that the basic meaning of a statement cannot be reduced to what a speaker does in using it, that there must be a 'locutionary' speech-act prior to the illocutionary one. John Searle, for example, has spoken of all such analyses as committing the 'speech act fallacy'.[11] But why is it a fallacy? The fact that in many cases we have a locutionary speech act and a meaning composed of sense and reference or connotation and denotation, which remain stable while the sentence is put to various uses, while various illocutionary acts are performed, does not entail that this is always so. Indeed it may clearly not be so. The one-word sentence 'Objection!' may have as its established conventional meaning that it is used to make a protest, and nothing more. The argument must then be that while sentences might have a purely illocutionary meaning, there is something about the way in which those that predicate 'good' or 'true' or 'probable' and so on are used which shows that they are not like this. In particular, it is argued that they can be embedded as clauses in more complex sentences, and that in these embedded positions they cannot be performing the speech act that it is supposed to be their very meaning that they perform, and yet that in their embedded positions they must still have the same meaning as when they are sentences on their own. For example in 'If he is a good man, God will protect him; he is a good man; therefore God will protect him', 'He is a good man' must have the same meaning in both places, or the *modus ponens* argument would be invalid; but it cannot be being used to commend him where it occurs within the antecedent of the conditional, but must have some other meaning there; consequently its meaning when it occurs as a sentence on its own, as the second premiss, cannot be simply that it is being used to commend him.

The argument is not as conclusive as it seems to be. R. M. Hare has shown that it is possible to explain the meaning of an

[11] J. R. Searle, *Speech Acts*, pp. 136–41; cf. P. T. Geach, 'Ascriptivism', *Philosophical Review*, 69 (1960), 221–5.

embedded good-sentence as a natural development of the commendatory meaning of a simple good-sentence in such a way that the *modus ponens* argument and similar constructions still work as expected.[12] In any case, this argument is two-edged when applied to 'if': for as we saw it is distinctly awkward to embed an if-sentence within the *antecedent* of another conditional: the form 'If if *P*, *Q*, then *R*' is not much used outside logic textbooks.[13] But other embeddings are certainly admissible. 'If *P* then if *Q* then *R*' is quite all right. But this lends itself easily enough to the suppositional analysis. The speaker supposes that *P*; makes the further supposition that *Q* within the scope of the first; and asserts *R* within the scope of both. There is nothing in the included position of the clause 'if *Q* then *R*' to stop it from being used for the complex speech-act of asserting *R* within the scope of the supposition that *Q*. Another sort of embedding is indirect speech. But the form 'He said that if *P*, *Q*' proves nothing about the meaning of 'If *P*, *Q*'. If the latter expresses the complex speech-act suggested, then 'He said that if *P*, *Q*' merely reports that he performed that speech-act. 'He believes that if *P*, *Q*' might seem more decisive: it requires that 'If *P*, *Q*' should express something that can be the object of *belief*, and the content of a complex speech-act can hardly be that. 'He believes that suppose that *P*, within the scope of that supposition, *Q*' does not make sense: it is not merely grammatically faulty, there is no coherent way of putting these components together. The only possible defence is to say that 'believes' here changes its meaning: 'He believes that if *P*, *Q*' might mean 'He is in such a state of mind that he is prepared to assert *Q* within the scope of *P*'. But I admit that this is a bit strained. Similarly, what about 'It is true that if *P*, *Q*'? It is hard to see how on the supposition account, any more than on the condensed argument account, 'If *P*, *Q*' could be strictly true, and yet here just this is being said by a perfectly standard sentence-form.

This kind of evidence, plus the very fact that the condensed

[12] 'Meaning and Speech Acts', *Philosophical Review*, 79 (1970), 3–24.

[13] It is, indeed, easier to embed a conditional in the antecedent of a since-statement: 'Since if *P*, *Q*, *R*' is quite an acceptable form. This *may* reveal a propositional employment of the if-sentence, but even this can be construed otherwise, as 'Suppose that *P*; (within the scope of that supposition) *Q*; in view of all this, *R*'. That is, the carrying out of the supposition-assertion performance can itself be used as the reason that supports a further assertion.

argument and supposition accounts are prima facie strange and have not been immediately attractive to most philosophers, compels us, I think, to make some concession. One possibility is to say that the conditional form functions initially as an abbreviation of the supposition-assertion-within-its-scope pattern, but then is treated as if it made some statement: it is mistaken for a statement-making form and operated on accordingly, but there is a conflict between this mode of external treatment and the internal content which as a conditional it can't help having. A milder and more plausible suggestion would be this: granted that there are the above-mentioned external pressures which are tending to force the conditional form to take on a statement-making role—arising essentially from the condensation or telescoping of the supposition-assertion speech-act which it performs—we may admit that the conditional form sometimes gives way to those pressures and takes on a statement-making role. In Peter Long's terminology, any if-sentence may have both a propositional and a non-propositional employment.[14] The non-propositional employment is, as Long says, primary, and the suppositional account explains it. The propositional employment is secondary, and comes into play when the condensation of the previous suppositional procedure (most frequently some inference) reaches the point where we tend to treat the product as a single unit. But then, *what* proposition does the if-sentence express? It seems to me that there are several possibilities, and we make use sometimes of one, sometimes of another. An if-sentence in its propositional employment may sometimes assert the corresponding material conditional; it may sometimes assert what the metalinguistic account says it does; or sometimes the speaker and/or hearer may take the notion of possible worlds literally and concretely, and treat the conditional as describing some actual state of affairs in a world that is not, or may not be, our own. There is no need to be exclusive here.

This concession and the resulting secondary readmission of accounts which we had criticized and rejected as general accounts of the meaning of if-sentences does take away from

[14] P. Long, 'Conditional Assertion', *Aristotelian Society Supplementary Volume*, 45 (1971), 141–7. The article by D. Holdcroft in the same symposium, 123–39, includes useful criticisms of von Wright.

the unity of our theory. But not too much. We still have a unitary theory of the core-meaning of if-sentences, and the secondary meaning is readily explained as a natural development of the primary one.

Our discussion of conditionals can be related to the account in Chapter 1 of different kinds of analysis. The suppositional account is an example of what I called, towards the end of that chapter, an external description. But would any other sort of analysis be available for if-sentences? We are not concerned with helping those whose native language is not English to use and understand if-sentences, but of course this could be done, most obviously by using the equivalent forms in the foreigner's native language, but conceivably by using some English synonymous or nearly synonymous construction like 'Suppose that P; in that case Q'. This *might* work; someone with an incomplete knowledge of English might know how to use 'Suppose that . . . in that case' without knowing how to use 'If . . . then . . .' but I would stress that this is not the point of my suppositional analysis. It is an attempt at an external description, not at a nearly synonymous translation. An analysis of the kind that picks out the perceptual cues involved in a habit of recognition would not be in general applicable to conditionals; it is pretty clear that they are not in general reports of anything that is even fairly directly perceived. Curiously enough, the if-sentences that come nearest to having perceptual cues associated with them are the counterfactuals latent in the elementary recognition of sequences as causal: 'If I hadn't touched that, my finger wouldn't be hurting now.'

We have more chance of finding an implicit logic or depth grammar of if-sentences. If the materialist analysis had been satisfactory, it would have been as an analysis of this kind. Moreover, it would have been an example of the sort that provided one of our ways of solving the paradox of analysis; it would have been a case where we used a form of expression with a certain implicit logic but without in general realizing that that was its logic. But as I have said this would conflict with a psychological interpretation of Grice's defence of the materialist view. On the other hand, the simpler form of the possible worlds—or rather possible situations—account is defensible both as a logical analysis and as an account of what speakers standardly intend to convey: the open conditional 'If *P*, *Q*' can be expanded into 'In the possible situation where *P*'—or 'that *P*'—'(which may or may not be actual), *Q* also', and the corresponding counterfactual into 'In the possible but not actual situation where *P*'—or 'that *P*'—'*Q* also'. As expansions, these seem correct, and the relation between the two explains the easily confirmed thesis that someone who asserts an open conditional and then finds that its antecedent is unfulfilled is committed to asserting the corresponding counterfactual and cannot simply claim 'So I was right'. My dissatisfaction with this simpler account was merely that a further, external, analysis of talk about possible situations is called for, while I rejected the attempt to use an elaboration from possible situations to systems of possible worlds to supply truth-conditions for all conditionals.

Let us reconsider the question whether all, or any, conditionals have truth-values. Is there any reason why they should? One possible answer is that only then could the structural semantics of these forms be incorporated in a Tarskian or Davidsonian truth-definition for the language. Admittedly if conditionals did have precise truth-conditions, their meaning could be neatly presented in this way. But the argument is equally forceful in reverse. If their meaning is best explained by an account which either does not allow or does not require them to have truth-values, this will show that the construction of a truth-definition is to that extent an inadequate presentation of the meanings of sentence structures in a natural language. But another possible answer is that people do frequently ascribe truth and falsehood both to if-sentences themselves and to kinds of statement (e.g. causal and dispositional ones) whose logical analysis seems to involve if-sentences. If such sentences could not be used to make true statements, we should have to admit that most of us are badly wrong about this aspect of the logic of our language.

There are two main arguments on the other side, that is, for denying that conditionals in their basic use have truth-values. The point is not that the complex speech act of supposing and asserting within the scope of that supposition could not itself have a truth-value; no more could the simpler speech act of asserting, and yet what is asserted does have a truth-value. It is rather that when we move from an external description to an expansion of the meaning of if-sentences as seen from the user's point of view, we still seem not to have something susceptible of truth. 'In the possible situation that *P*, *Q* also', just because it purports to describe a merely possible situation, does not leave room for things to be, or not to be, as is stated. Simple truth, truth in the strict sense, can belong only to descriptions of what is actual. The second argument is that a counterfactual (at least so long as the speaker is right in believing the antecedent to be unfulfilled) lacks determinate truth-conditions; we cannot retain, to conjoin with its unfulfilled antecedent, all features of the actual world. There is, then, an unresolvable arbitrariness about what we retain and what we abandon, and it may well be that the consequent would follow from one selection of retained features whereas its denial would follow from another. I have argued that the elaborate possible worlds account cannot remove this arbitrariness but only relocates it.

But some compromise may be possible. First, what are we to say about counterfactuals whose antecedents are stated so fully that no arbitrariness remains? Where '*P*' is unfulfilled, but '*P* & *X* & *Y* & *Z*' *entails* '*Q*', can we not say that the counterfactual 'If *P* & *X* & *Y* & *Z* then *Q*' is true? No reason in terms of arbitrariness can now be given for not doing so. And equally, where '*P* & *X* & *Y* & *Z*' entails 'not-*Q*', can we not say that the counterfactual 'If *P* & *X* & *Y* & *Z* then *Q*' is false? What logically must hold for all situations can be said to hold for possible but not actual ones. I think this can be conceded, though it demands some extension of the simple notion of truth. What is simply true here is that the entailment holds, that '*P* & *X* & *Y* & *Z*' entails '*Q*' (or 'not-*Q*', as the case may be), and it is by an extension that we allow what amounts to a description of a possible situation to be called true (or false). In any case, most counterfactuals are not of this kind, their antecedents do not entail either their consequents or the negations of their consequents, and for them the problem of arbitrariness remains.

Secondly, it is plausible to say (with, e.g., von Wright) that a conditional whose antecedent is in fact fulfilled—whether it has been expressed as open or as counterfactual—is true if its consequent is also realized, and false if it is not. That is, non-material conditionals obey the top two lines of the standard truth table, though not the others. We can, on our principles, explain why this is plausible. If someone says something of which 'In the possible situation that *P*, *Q* also' is an equivalent expansion, then if '*P*' is fulfilled, we can identify the-possible-situation-that-*P* with the actual situation: he has in effect said that '*Q*' holds of the actual situation, so his statement will have the same truth-value as '*Q*'. This is most obvious if we use the expansion suggested above, 'In the possible situation that *P* (which may or may not be actual), *Q* also'. This point overrides the argument that because a conditional purports to describe a merely possible situation, it cannot be simply true. A conditional (other than a 'factual conditional', i.e. a since-statement) introduces a situation merely *as* possible, but if it is an open conditional it does not preclude its being actual as well, and although a counterfactual does preclude this, it may be mistaken in doing so: we may then say that the implied or suggested denial of the antecedent is false, but that the conditional part of the counterfactual still has the same truth-value as the consequent.

This is plausible, but let us test it further by considering four cases. All four begin with a father saying to a child 'If you poke your finger into that monkey's cage, you will get it nipped off' and with the child poking its finger into the cage none the less. The first case continues with the monkey snarling, snapping at the finger, and biting it off. In the second case the monkey remains placid, the child withdraws its finger, but at that moment a large bird swoops down and nips off the finger. In the third case the monkey remains placid, nothing else intrudes, and the finger remains whole. In the fourth case the monkey snarls, snaps at the finger, but just at that moment a large object falls from the roof of the cage and deflects the monkey's attack, and again the finger survives intact. In which, if any, of the four cases is the father's statement true? (We want, of course, to confine our attention to the truth or falsehood of what the father *said*, as opposed both to what he may have suggested or had in mind and also to his reasons for saying it.) It would be very hard to deny that the father's statement was false in the third case. As long as we stand firm on attending to what was said, we must say the same about the fourth: there the father would presumably withdraw to claiming that all he *meant* was that there was a serious risk of the finger's being bitten off, and this has been thoroughly confirmed. Similarly, it is very hard not to admit that the statement is true in the first case; if the father here said 'I told you so' we might condemn his harshness but we could not deny his claim. A consequentialist would say that the statement was true only in this first case and not also in the second, but I think he would be wrong: all the father said was how things would be on a certain supposition; the actual course of events has realized that supposition, and things have turned out as it was said that they would.

On the other hand, if we consider a fifth case, in which the child does not poke its finger in, and the finger remains whole, and a sixth in which the child does not poke its finger in, but the bird nips it off, we must say that the father's statement is neither true nor false in each of these. It is not true as a material conditional would be in both; it is not false in the sixth case, for the father did not say, though he may have suggested, that the finger would not be nipped off if it were not poked into the cage; and however reasonable, acceptable, and so on both the open conditional and the counterfactual 'If you had poked it in it would have been nipped off' may be made by, say, the sad fate of another child's finger a moment later, neither the open conditional nor the counterfactual is in this case strictly true.

It is, then, not merely compatible with the suppositional account but a consequence of it that conditionals whose antecedents are fulfilled are all either true or false in a strict sense, provided only that their consequents have truth-values; also, by a natural extension we can allow truth or falsehood to be ascribed to conditionals whose antecedents are unfulfilled but entail their consequents or the negations of their consequents; but we can insist that other conditionals whose antecedents are unfulfilled are not susceptible of truth or falsehood. The general semantic structure of conditionals does not provide them with truth-values in all cases, but in some only; and this would be at least an awkward thing to accommodate in an attempt to display the whole structural semantics of a language in a truth-definition.

§ 10. *The Logic of Conditionals*

We can use the suppositional account to clear up some problems about the logic of conditionals and their contrasting relations with natural laws and accidental generalizations.

Can conditionals be contraposed? Is 'If *P*, *Q*' equivalent to 'If not-*Q*, not-*P*'? We should expect it to be. But we are also inclined to say that 'If *P*, *Q*' is compatible with 'If not-*P*, *Q*'—e.g. 'If Boycott makes a century, England will win' is compatible with 'If Boycott doesn't make a century, England will win'; there is even a standard form for the conjunction of two such conditionals: 'England will win whether Boycott makes a century or not.' On the other hand, 'If *P*, *Q*' seems not to be compatible with what I shall call the contrary conditional, 'If *P*, not-*Q*'—e.g. 'If you go to Benidorm you'll have a lovely time' is not compatible with 'If you go to Benidorm you won't have a lovely time'. In general the point of saying 'If *P*, *Q*' is to make a discrimination, to say that on the supposition that *P*, we shall have *Q* and we shall not have not-*Q*. But we cannot retain, without qualification, the compatibility of 'If *P*, *Q*' and 'If not-*P*, *Q*', the incompatibility of 'If *P*, *Q*' and 'If *P*, not-*Q*', and equivalent contraposition all together. For if 'If *P*, *Q*' and 'If not-*P*, *Q*' were compatible, so would their contrapositives 'If not-*Q*, not-*P*' and 'If not-*Q*, not not-*P*' be compatible; and these two are of the form 'If *P*, *Q*' and 'If *P*, not-*Q*' (with 'not-*Q*' replacing '*P*' and 'not-*P*' replacing '*Q*').

This problem does not, of course, arise with material conditionals, because '$P \supset Q$' and '$P \supset -Q$' are compatible; they merely together entail '$-P$'.

What, then, are we to give up? This will depend on the purposes for which we are using the conditionals. Granted that every (primary) use of a conditional is tantamount to asserting something within the scope of some supposition, there may be different reasons for doing so, different illocutionary acts involved. If one is treating what is supposed as a genuine possibility—or in counterfactual cases as having been a genuine possibility—that is, as something not ruled out by the background assumptions in the light of which one is considering it, but as something which, in view of those background assumptions, would have—or would have had—some determinate outcome, then one will regard 'If *P*, *Q*' and 'If *P*, not-*Q*' as incompatible; they are not incompatible in themselves, as conditionals, but together they are incompatible with this which we may call the *straightforward* or *direct* use of conditionals. Alternatively, one may treat what is supposed as a possibility only in a weaker sense, as something which one's background assumptions rule out, but which can nevertheless be considered initially as possible with a view to showing, in the end, that it is impossible: in other words, there is an *indirect* or *reductio ad absurdum* way of using a supposition. In this sort of use, 'If *P*, *Q*' and 'If *P*, not-*Q*' are compatible: they do not conflict because, on the assumptions with respect to which one is considering possibilities, it is not really possible that *P*.

Someone who asserts a pair of conditionals of the form 'If *P*, *Q*' and 'If not-*P*, *Q*' (in their direct use) will be relying implicitly on some set of background assumptions '*S*' such that '*S*' and '*P*' together entail '*Q*' and equally '*S*' and 'not-*P*' together entail '*Q*'. But then '*S*' itself must entail '*Q*', and rule out 'not-*Q*'. Consequently no conditional of the form 'If not-*Q* . . .' will be acceptable to such a speaker for direct use, and in particular he cannot accept for direct use the contrapositives of 'If *P*, *Q*' and 'If not-*P*, *Q*'. But he can accept them for *reductio ad absurdum* use; he can assert both 'If not-*Q*, not-*P*' and 'If not-*Q*, not not-*P*' as a way of bringing out the impossibility of 'not-*Q*' on the assumptions on which he is implicitly relying.

Thus the incompatibility of 'If *P*, *Q*' and 'If *P*, not-*Q*' holds for the direct use, and contraposition is accordingly restricted. It is not done away with even for the direct use. There will be many direct uses of the form 'If *P*, *Q*' such that 'not-*Q*' is not ruled out by the background assumptions, conditionals of the form 'If not-*Q* . . .' will therefore be acceptable for direct use, and then using 'If *P*, *Q*' will commit one to being prepared to use 'If not-*Q*, not-*P*'. As long as the antecedent 'If not-*Q*' is admissible, the contraposition is valid. For the indirect use, contraposition holds without restriction, and 'If *P*, *Q*' and 'If *P*, not-*Q*' are compatible.

We can easily illustrate these general principles. Anyone who says 'If Boycott makes a century, England will win', implicitly relying on assumptions which do not in themselves ensure that

England will win without the premiss 'Boycott makes a century' —so that he is *not* prepared to say also 'If Boycott doesn't make a century, England will win'—will allow the real possibility of England's not winning and is logically committed to the contrapositive 'If England does not win, Boycott will not have made a century'. But someone who says both 'If Boycott makes a century, England will win' and 'If Boycott doesn't make a century, England will win' cannot coherently consider it as a possibility, in relation to whatever background assumptions he is for the moment relying on, that England should not win, and cannot therefore use directly either contrapositive, and he is not logically committed to either (since an indirect use would be unnatural in this context). If this speaker does begin a statement with the words 'If England doesn't win . . .' he must be changing his ground, moving to some different background assumptions, and then there will be no simple logical connections with what he said in reliance on the previous assumptions. On the other hand, the contrary conditionals 'If there were an infallible perception it would be instantaneous (in order to be infallible)' and 'If there were an infallible perception it would not be instantaneous (being a perception)' are compatible if they are used together in a *reductio ad absurdum* argument to show that there could not be an infallible perception.

Such a pair of compatible contraries will naturally be expressed in a subjunctive or counterfactual form, since they are compatible only because the speaker is committed to rejecting their common antecedent. But we must not convert this rule, and say that pairs of contrary subjunctive and counterfactual conditionals are always compatible. They are not. Subjunctive and counterfactual conditionals are normally used directly, just as open conditionals are. 'If you had struck that match it would have lit' and 'If you had struck that match it would not have lit' are just as incompatible, in their ordinary direct use, as are 'If you strike that match it will light' and 'If you strike that match it will not light'.

We should not confuse the two distinct points that a pair of contrary counterfactuals (i) will be compatible in an indirect use, and (ii) may be separately acceptable, with a shift of assumptions, in a direct use. 'If you had struck that match, it would not have lit (because it was wet)' and 'If you had struck that match, it would have lit (because you are a careful person and would have dried it first)' are separately acceptable for direct use. But since they rely on different beliefs retained for use in conjunction with the assumption, they cannot be conjoined for direct use to give 'If you had struck that match it would both have lit and not lit', which would be nonsense in a direct use. This illustrates (ii). But, to go back to (i), there will be an indirect use of these two counterfactuals together: 'That match was wet' and 'You are (in the required sense) a careful person' are together incompatible with the supposition 'You struck that match'; and this fact might be brought out by an indirect use of the conditional 'If you had struck that match it would both have lit and not lit'—for the reasons given above—which relies on the retaining of both the assumptions which are together incompatible with the supposition. The conclusion drawn from this double counterfactual is, of course, 'So you (as you are) could not have struck it (as it is)', not merely 'So you did not strike it', which is conceded automatically by the counterfactual form.

This shows how we can deal with the problem of competing counterfactuals. 'If Bizet and Verdi had been compatriots, Bizet would have been Italian' and 'If Bizet and Verdi had been compatriots, Verdi would have been French' seem equally plausible. But they do compete; if we try to combine them we get the absurdity that if they had been compatriots, Bizet would have been Italian and Verdi French. In the first place, these counterfactuals are separately acceptable. If, where we introduce the belief-contravening supposition that the two composers were compatriots we retain the true belief that Verdi was Italian, we shall assert, within the scope of that supposition, that Bizet was Italian too. If, instead, we retain the equally true belief that Bizet was French, we shall assert within the scope of our supposition that Verdi was French. The counterfactuals can clearly be interpreted as of the condensed argument sub-species: in each case the consequent would follow from the supposition in conjunction with a true premiss and certain linguistic rules about the term 'compatriots' and the nationality descriptions 'French' and 'Italian'. But the three premisses 'Verdi was Italian', 'Bizet was French', and 'Bizet and Verdi were compatriots' form, in the light of the linguistic rules, an inconsistent triad. There is therefore no *direct* use for the combined counterfactual, in which we should have to retain both the premisses 'Verdi was Italian' and 'Bizet was French' along with a supposition that conflicts with their conjunction. But there is still a possible *reductio ad absurdum* use. We could say 'If Bizet and Verdi had been compatriots Bizet would have been Italian and Verdi would have been French (for the reasons indicated) and so they would not have been compatriots; that is, they couldn't have been compatriots.' And of course they couldn't have been, so long as we are considering only possibilities which allow us to retain at once 'Verdi was Italian' and 'Bizet was French'. And this is all there is to it. Since we have denied that non-material conditionals can be true, the question which of these competing counterfactuals is true does not arise. Neither is true; each is acceptable in certain circumstances; but they are co-acceptable only in an indirect use. Since in ordinary circumstances we have no reason for preferring one to the other of these, we are not likely to be very strongly tempted to use either.

These counterfactuals were competing on equal terms, but a similar logical pattern applies even where symmetry is absent, e.g. where one of the competing counterfactuals, but not the other, is sustained by a causal law. The statements 'Cyanide is a deadly poison', 'Jones is alive', and 'Jones took cyanide' form an inconsistent triad, and the first and second of these sustain, respectively, the competing counterfactuals 'If Jones had taken cyanide, he would not be alive' and 'If Jones had taken cyanide, cyanide would not have been a deadly poison'. But these are not on equal terms: we are much more prepared, when introducing the belief-contravening supposition that Jones took cyanide, to stick to the law that cyanide is a deadly poison than to the particular fact that Jones is alive. The point is not that the former generalization is 'so secure that we are willing to retain it at all costs':[15] the fact that Jones is now alive may be equally 'secure'. The point is that the counterfactual form concedes that Jones did not in fact take cyanide, so that the supposition that he did take it introduces a *different* situation from the actual one, and there is no reason for taking the

[15] Cf. N. Rescher, op. cit., p. 198.

observation that Jones is alive in the actual situation as informing us about the different possible one. This is not because the law about cyanide is known or secure, but merely because we know that there are causal laws, that a difference in a temporal antecedent is often followed by a different outcome. That this is the point is confirmed by the fact that the *open* conditional 'If Jones took cyanide, cyanide is not a deadly poison' is quite natural and plausible. This is so because we can quite reasonably retain the observed fact that Jones is now alive for use along with the supposition, considered as an *open* possibility, that he took cyanide; we are now considering a situation consisting of the whole of the actual one along with the fulfilment of the antecedent, which is being treated as neither known to be fulfilled nor known not to be fulfilled. In the possible situation thus constructed, the law that cyanide is a deadly poison cannot hold. The corresponding counterfactual is not plausible because the contrary-to-fact supposition, just by being contrary-to-fact, introduces a situation other than the actual one, and so does away with our reason for retaining, within the scope of this supposition, such particular features of the actual situation as that Jones is now alive.

This brings us to what is perhaps the greatest benefit resulting from our fuller understanding of conditionals: the light that this throws upon their relations with causal laws and accidental generalizations and consequently upon the nature of causal laws themselves. Why do causal laws entail or sustain counterfactuals whereas accidental generalizations do not? Does this fact show, as is widely believed among philosophers, that causal law statements include, in their meaning, something stronger than merely factual universality? Do these statements implicitly assert the existence of some sort of 'natural necessity' in the events themselves? Is there some special virtue either in causal law statements, or in the objective laws which they report, which enables them to entail counterfactuals, mysterious truths that hold beyond the actual world and govern the realm of possibilities as well?

My contention is that this way of asking the questions is thoroughly misleading. Counterfactual conditionals are not to be taken literally as truths about possible worlds, but as a species of human procedure. They are just non-material conditionals plus a hint that their antecedents are unfulfilled, and non-material conditionals merely express the asserting of something within the scope of some supposition—which may be done for any one of a number of reasons, which may themselves be reasonable or unreasonable. All sorts of statements can sustain counterfactuals, including, as we have seen, such singular statements as 'Bizet was French'. The real problem is not to find any extra virtue in causal laws, but to find what special deficiency in accidental generalizations prevents them from sustaining counterfactuals. Or, more generally, to explain why some logically formulable counterfactuals are more acceptable than others.

Once we ask the right question it is comparatively easy to find the answer. Let us consider the accidental generalization 'Everyone in this room understands Italian', established, presumably, by complete enumeration, by checking each individual in turn. To use this to sustain the counterfactual 'If Mr. Chou En-Lai were in this room he would understand Italian' would be to introduce the supposition—admitted to be false—that Mr. Chou En-Lai is in this room, and then to assert, within its scope, that Mr. Chou En-Lai understands Italian, using the supposition and the enumeratively established universal together to yield this result. But it is not reasonable to use them together. Since our sole ground for believing this universal was the enumerative check, that ground collapses as soon as we add the supposition that someone *else* is in the room; someone who—as the counterfactual form concedes—is not in fact in the room and whose understanding of Italian has therefore not been checked by this enumeration. The adding of the contrary-to-fact supposition takes us from the actual situation to a different, merely possible, one, one in which we have not checked everyone's understanding of Italian. If the universal were true and Mr. Chou En-Lai were in the room then of course he would understand Italian; but since our reason for believing the universal evaporates as soon as we introduce the supposition, we cannot reasonably take this universal as we know it and this supposition as joint premisses in an argument, even a telescoped argument; and unless we do so we have no reason for asserting the counterfactual. Since the complete check was our only reason for believing the universal we are not justified in retaining it within the scope of our supposition, and in fact we are not prepared to do so.

This account is confirmed if we contrast the counterfactual with an open conditional. The accidental generalization 'Everyone in this room understands Italian' does sustain the *open* conditional 'If Mr. Chou En-Lai is in this room he understands Italian'—that is, one of the persons present may be Mr. Chou En-Lai disguised or unrecognized, and if so he has passed the test of his understanding of Italian. This is acceptable because the open supposition does not carry us to a different situation, but—like 'If Jones took cyanide' in our previous example—only *adds* to the actual situation an item that is taken as neither known to be so nor known not to be so. Thus the open supposition that Mr. Chou En-Lai is here does not cancel our reason for retaining the enumeratively established universal, whereas the counterfactual supposition—that is, the supposition that he is here coupled with the admission that he is not here—does cancel it. It is the contrary-to-factness of the antecedent that makes us unable to use an enumeratively established universal within its scope.

This account can easily be extended to cover examples where the accidental generalization is known not by a complete enumeration but by some other but logically equivalent process. If we know that none of the stones in this box is radioactive because a Geiger counter near by shows no response, this universal does not sustain the counterfactual 'If that other stone were in this box it would not be radioactive', again because the supposition that some *other* stone is in the box undermines the evidence of the Geiger counter as a reason for asserting the universal within the scope of the supposition.

On the other hand, a generalization sustains a counterfactual if our reason for adhering to it is such as to survive its being put within the scope of a belief-contravening supposition. Let us look at some contrasting examples before proceeding to a general explanation.

Suppose that the gathering in this room is a meeting of the Italian Poetry Circle, and this fact is clearly announced in a notice on the outside of the door in several languages, including Chinese. This would give pretty good grounds for saying that if Mr. Chou En-Lai had been here he would have understood

Italian. Still stronger grounds would be provided by the presence of a doorkeeper who had been instructed to let in only those who proved their understanding of Italian. Similarly if this box were, say, the left-hand box of a pair attached to a collecting and sorting device which pushes all radioactive objects it encounters into the right-hand box, knowledge of this device would sustain the counterfactual 'If that other stone were in this box it would not be radioactive'.

Lying behind the grounds relied upon in these cases are, of course, causal laws: in the first example we have devices which (more or less efficiently) cause the exclusion of those who do not understand Italian, and in the second a device which causes the exclusion of radioactive objects. So the question is, why do causal laws work in the way they do? That is, why can we (i) combine a law with suppositions that go beyond cases for which the law has been checked, and so advance open or subjunctive conditionals, and (ii) combine it with suppositions which we take to be not fulfilled, which we regard as altering the extension of the law's subject term, and so advance counterfactual conditionals? The answer is simply, because we have what we take to be good inductive evidence for the law.

If we have good inductive evidence for the law 'All *A*s are *B*', then this evidence supports the conclusion that an unobserved *A* is *B*; it therefore justifies an argument from the supposition that a certain object *X* is an *A* to the conclusion that *X* is *B*; it therefore justifies us in asserting that *X* is *B* within the scope of the supposition that *X* is an *A*, and hence for saying that if *X* is an *A*, it is *B*. Such evidence will, therefore, sustain the open conditional 'If *X* is an *A*, it is *B*'. But this evidence is logically related in exactly the same way to the argument from the supposition that *Y* is an *A* to the conclusion that *Y* is *B*, even if we happen to know or believe that *Y* is not an *A*; it therefore justifies us in asserting that *Y* is *B*, within the scope of the contrary-to-fact supposition that *Y* is an *A*, and hence for saying that if *Y* had been an *A*, it would have been *B*. Such evidence will therefore sustain also the counterfactual 'If *Y* had been an *A*, it would have been *B*'. Formally, all that is required to let a law sustain counterfactuals is that there should be the same logical relation (i) between the evidence and the proposed law (covering unobserved instances) as things are, and (ii) between the evidence and the proposed law with things otherwise the same but with additional instances of the law's subject term. And this holds for all ordinary inductive reasoning.

To enable a law to sustain counterfactuals, then, *all* that is needed is that it should be supported by what we take to be good inductive evidence. It is no part of my task, in offering this explanation, to say either what is good inductive evidence or why it is so. The hard fact is that we do reason inductively: given that we do, the sustaining of counterfactuals by laws which are (directly or indirectly) supported inductively is an automatic consequence, in view of the general account of conditionals and counterfactuals that I have offered. Inductive evidence is, by definition, projective; inductive evidence for a law provides a ground for believing that law which is not impaired either (i) by the supposition that there is an instance of the subject term which has not been included in the evidence, or (ii) by the supposition that there are additional (contrary-to-fact) instances of the subject term. The sustaining of counterfactuals by laws or 'nomic universals' is nothing more than the projective force of inductive evidence in a new guise. This explanation has not made use of any notion of a special *content* in a causal law; unless it can be shown that the inductive reasoning itself requires a special content in its conclusions—and there is no plausibility in this suggestion—there is nothing in the sustaining of counterfactuals to show that causal laws have a content or meaning that in any way differs from that of a straightforward factual universal. Causal laws *may* also contain certain special sorts of information, but there is nothing in their distinctive relation to counterfactuals to require that they should. Their sustaining of counterfactuals is exhaustively accounted for, not by their content, but by the inductive character of the evidence which directly or indirectly confirms them. In so far as the term 'nomic universal' suggests a special content, it should be discarded forthwith.

I conclude, then, that we need no longer lament that we do not know how to construe conditionals or that the exact analysis of counterfactuals in particular is an unsolved riddle. The suppositional account does satisfactorily elucidate all the standard uses of conditionals—peripheral as well as central—

and also relates them intelligibly to other uses of the word 'if'. Puzzles about the logic of conditionals can then be resolved, and the power of causal law statements to sustain counterfactuals loses at once its mystery and its supposed profound significance.

4

DISPOSITIONS AND POWERS[1]

§ 1. *The Meaning of Dispositional Statements*

MANY of the ways in which we describe ourselves and the physical world involve dispositional predicates. We say that things are malleable or impermeable, that people are generous or touchy or inconsiderate. Some predicates, like these, are obviously dispositional; but it has been argued that many that are not immediately recognized as dispositional are so too. That this is true of many terms that would be said to describe mental states or processes is a central doctrine of Ryle's *Concept of Mind*. Sir Karl Popper, in suggesting that the propensity interpretation of probability may 'offer a new metaphysical interpretation of physics (and incidentally also of biology and psychology)' remarks that 'we can say that all physical (and psychological) properties are dispositional'.[2] Similarly Goodman says that 'almost every predicate commonly thought of as describing a lasting objective characteristic of a thing is as much a dispositional predicate as any other'.[3] More recently Harré has revived the old-fashioned term 'powers' and argues that 'the concept of "power" can play a central role in a metaphysical theory appropriate to a realist philosophy of science',[4] and his powers are either identical with dispositions or closely related to them.

But it is not clear how dispositional predicates are to be understood, or exactly what we are saying about a property if we classify it as dispositional. Goodman's remark suggests that

[1] This chapter is based on a symposium on 'Probability and Dispositions' with Dr. D. H. Mellor, at a meeting of the British Society for the Philosophy of Science in September 1971. While Dr. Mellor and I are still in radical disagreement on some issues, a number of the points I make are due directly to him, and my presentation of many others has been, I hope, clarified by discussion with him.

[2] *Observation and Interpretation in the Philosophy of Physics*, ed. S. Körner, p. 70.

[3] *Fact, Fiction, and Forecast*, pp. 44–5.

[4] 'Powers', *British Journal for the Philosophy of Science*, 21 (1970), 81–101, esp. 81.

such predicates are only *thought of* as describing lasting objective characteristics, and that in detecting their dispositional character we shall correct this belief; but Harré's powers are relatively permanent differences in intrinsic nature.[5] There are, in fact, conflicts between 'phenomenalist' and 'realist' views about dispositions, and the realists are not all realistic about the same sorts of item. Besides, dispositions and powers are disturbingly reminiscent of faculties, dormitive virtues, and the like, and it is widely believed that there is something fundamentally wrong with concepts of this group, though few would be able to say exactly what is wrong with them. There are, then, problems both about the meaning of dispositional terms and about the nature and status of dispositional properties. It is generally agreed, however, that dispositions have something to do with conditional statements, and particularly with counterfactual conditionals, so that we may be helped in sorting out these problems by some of the results of our discussion of conditionals.

There are several kinds of dispositional terms, though perhaps not as many as is suggested by those who are inclined to call dispositional all predicates that are non-occurrent or (what is not the same thing) all that are non-manifest. Ordinary language applies the word 'disposition' mainly to human characteristics like generosity or touchiness rather than to such physical features as malleability or impermeability, and the latter might be more naturally called dispositional properties. The former group includes mainly what have been called active powers, which are concerned with what the person or thing that has them will do or tends to do in certain circumstances, while the latter group includes mainly passive powers, which are concerned with what can (or cannot) be done to the thing or with how it is likely to be changed by something that happens to it. But the most fundamental problems extend right across both these dividing lines, and in general I shall ignore them, and I shall use the terms 'disposition' and 'dispositional property' interchangeably. A distinction which it is more useful to draw, because it simplifies the discussion, is between sure-fire dispositions and probabilistic ones. A man's irritability would be a sure-fire disposition if he was certain to become angry in response to a disturbance or provocation of more than

[5] Op. cit., pp. 85, 92.

a certain magnitude; it would be a probabilistic disposition if he was only likely to do so. I shall discuss dispositions first on the assumption that we are dealing with sure-fire ones; after some of the main problems have been settled, I shall relax this simplifying restriction. Another distinction is between dispositions that are manifested in just one way and those that are manifested in several different ways: fragility is manifested only by the object's breaking upon some sort of impact, but a high temperature is manifested by the thing's feeling hot, by the response of a thermometer in contact with it, and perhaps in other ways as well; if someone's believing that such-and-such is a dispositional property, it is manifested by his verbal behaviour in answering questions, by what he does to satisfy his wants, and so on. This distinction is not a very sharp one, since by choosing a broad enough description of the manifestation we may be able to make almost every disposition a singly-manifested one, and by choosing a set of very precise descriptions of the manifestations we can make every disposition a multiply-manifested one. Still, a rough distinction of this sort has some plausibility, and it matters in that the kind of analysis that is appropriate for singly-manifested dispositions may not be immediately extendable to multiply-manifested ones. Again for simplicity I shall begin with singly-manifested dispositions and consider the extension later.

The manifestations or displays of dispositional properties commonly involve causal or quasi-causal interaction between two or more things: sugar dissolves in water, a lump of lead is flattened by being hit with a hammer, one man becomes annoyed because another criticizes him or his work, or because a radiogram is played loudly next door, nitric acid corrodes copper. But it is of some interest that in such cases a disposition is frequently ascribed to just one of the interacting things: it is the sugar that is soluble, the lead that is malleable, the man who is touchy or irritable, the acid that is corrosive. Dispositions are to this extent treated as if they were intrinsic properties.

Dispositions are also treated as if they were relatively lasting properties which actually belong to things throughout the periods for which they are ascribed to them. A glass is fragile for a long time while it is not being knocked and broken, but also while it is being knocked and is breaking. A piece of lead is malleable both while it is and while it is not being hammered. A man may be irritable during periods when, thanks to the absence of provocation, he remains calm. But dispositional terms can and for clarity should be distinguished from predictive and retrodictive ones. To say that someone is a murderer is (usually) to say merely that he has committed at least one murder; but to say that he is murderous is to ascribe a disposition which he may never manifest by actually murdering anyone. There is a big difference between being a future Prime Minister and being a potential Prime Minister. To call someone a cigarette-smoker is to say that he has smoked a lot of cigarettes, but it is probably also to ascribe to him a disposition, a tendency to smoke cigarettes at frequent intervals; this, then, is a term which combines a retrodictive with a dispositional meaning, but the two elements can still be distinguished. Some terms have meanings that are undecided between the dispositional and the predictive. It is tempting to say that '... is mortal' just means '... will at some time be dead'; but the Greek myth according to which Heracles started off as mortal, but when he was about to die was changed into a god and never died at all, requires that 'mortal' should have the dispositional meaning 'liable to die'.

Discussion of the problem of dispositions, as of other topics, has often been obstructed by the confounding of three different sorts of question: questions about meaning, about what is being said when dispositions or powers are ascribed to things; questions about what we know, about when and how we are able to ascribe dispositions or powers; and questions about what is there, about what sorts of properties or states or processes are objectively present in the things that we describe in dispositional terms. It will be important to keep these questions apart as far as possible, and at any rate to be clear which kind of question any account that we propose is intended to answer. I shall concentrate first on questions of meaning.

Having distinguished dispositional statements from simple predictions, we might try to construe them as conditional predictions; for example, '*x* is soluble (in water)' might be taken to mean 'If *x* is placed in water it will dissolve'. However, since *x* might be insoluble now, and later become soluble, we cannot allow that 'If x is placed in water tomorrow it will dissolve' entails 'x is soluble now'; and conversely, since x might be soluble now but become insoluble later, 'x is soluble now' does not entail 'If x is placed in water tomorrow it will dissolve'. Though relatively lasting, dispositions are ascribed to things at times, not to continuing individuals absolutely. We need to say rather that 'x is soluble at t' means 'if x is placed in water at t it dissolves at t'. (Strictly, 'at $t+\delta t$' or 'a little later', but we can ignore this refinement.) But it will not do to interpret this as a material conditional. Carnap[6] once considered the suggestion that 'x is soluble'—'$S(x)$'—should be defined thus:

$$S(x) = (t)[W(x, t) \supset D(x, t)]$$

with '$W(x, t)$' meaning 'x is put into water at t' and '$D(x, t)$' meaning 'x dissolves at t'. But, as Carnap says, this fails because it would make soluble every object that was never put into water. And if '$S(x, t)$' is read as 'x is soluble at t',

$$S(x, t) = W(x, t) \supset D(x, t)$$

would fail similarly: it would make soluble at t every object that was not put in water at t. Carnap's more plausible suggestion that 'soluble' can be introduced by the analytic reduction sentence

$$(x)(t)[W(x, t) \supset (S(x) \equiv D(x, t))]$$

still fails to capture the ordinary meaning of 'soluble': it makes soluble (absolutely) anything which is ever put in water and dissolves, and insoluble (absolutely) anything which is ever put into water and fails to dissolve, so that something which was insoluble at t_1, and tested then, and which became soluble by t_2, and was tested then also, would be both soluble and insoluble absolutely. Correcting for this, we might try

$$(x)(t)[W(x, t) \supset (S(x, t) \equiv D(x, t))].$$

But though the meaning of 'soluble' is indeed such as to make this analytic, it leaves indeterminate the solubility or otherwise at t of any object that is not put in water at t, just as the previous reduction sentence leaves indeterminate the (absolute) solubility of anything which is never put in water.

[6] 'Testability and Meaning', *Readings in the Philosophy of Science*, eds. Feigl and Brodbeck, pp. 52–3 (reprinted from *Philosophy of Science*, 3 (1936)). I have used Storer's symbols (see below) instead of Carnap's own.

A more elaborate suggestion has been put forward by Storer.[7] He first defines a predicate 'WD' which applies to all and only those things which at some time dissolve in water, and a predicate '$\overline{WD}$' which applies to all and only those things which are at some time put into water but never dissolve. He then defines 'x is soluble' thus:

$$S(x) = WD(x). \vee. (\exists F)[F(x) . (\exists y)\{F(y) . WD(y)\}. \\ \sim (\exists y)\{F(y) . \overline{WD}(y)\}].$$

In effect, something is soluble if *either* it at some time dissolves *or* it has some property such that some things with that property are at some time put into water and dissolve and nothing with that property is at any time put into water without dissolving. This would still need some patching to provide for the possibility that something may be soluble at one time but not at another: we still need to define not '$S(x)$' but '$S(x, t)$'. More seriously, any definition of this sort might by accident produce bizarre and incorrect results. There might be a property F such that everything which had F and was put in water dissolved, and yet which had causally nothing to do with dissolving. For example, F might be the complex property of being cubical, transparent, and having been watched under sodium light for exactly four minutes by an albino mouse; it might be that only two things have F: this grain of sugar and that glass cube. Then if the grain of sugar is put into water and dissolves, but the glass cube is never put into water, the proposed definition makes this glass cube soluble. This objection trades, of course, on the fact that a simple regularity analysis of causal statements fails, while what Storer needs is an F which is causally sufficient for dissolving in water. More generally, this objection illustrates the point that for the analysis of dispositional statements we need conditionals which are not themselves purely material conditionals.

Having seen these difficulties, we may be tempted to equate dispositional statements with counterfactual conditionals: to say that x is now soluble is to say that if x had just now been put into water it would now be dissolving. But if we insisted on *counterfactuals*, we should get the absurd result that sugar ceases

[7] 'On defining "soluble" ', *Analysis*, 11 (1950–1), 134–7.

to be soluble when it is put into water, and that anyone who actually loses his temper is no longer irritable.[8] It is true that we have most use for dispositional descriptions when the corresponding episodes are not occurring: we have less need to call someone merely irritable if we can say that he is actually annoyed. But I do not think that the meaning of the dispositional terms requires that they should be withdrawn when the dispositions are being manifested: they are merely liable to be pushed aside by something better.[9] The counterfactual analysis thus has the opposite defect to that of some material conditional analyses: they yield undesired results for things that are not presently dissolving; the counterfactual yields undesired results for things that are.

There is, however, an easy way of escape from this dilemma: dispositional statements are equivalent to non-material conditionals, not specifically to counterfactual ones. We can analyse '*x* is soluble at *t*' as 'If *x* is put into water at *t*, it dissolves', where this 'if' is non-material; whether the statement would take an open, subjunctive, or counterfactual form depends on the speaker's beliefs about what happened at *t*. Changing the example, if I believe that glass *a* was not knocked at *t*, then I shall be prepared to say that it was fragile at *t* if and only if I am prepared to use the counterfactual 'If *a* had been (suitably) knocked at *t* it would have broken'. If I know that *a* was made a week ago, but have not heard what has happened to it since then, I shall be prepared to say that it was fragile yesterday if and only if I am prepared to use the open conditional 'If *a* was knocked yesterday it broke'; similarly I am prepared to say that it will be fragile tomorrow if and only if I am prepared to use either the open conditional 'If *a* is knocked tomorrow it will break' or, if I think it unlikely to be knocked, the subjunctive conditional 'If *a* were knocked tomorrow it would break'. (I am here treating fragility as a sure-fire disposition, and I am using '(suitably) knocked' as a broad term to cover any impacts of any of the sorts we should ordinarily regard as sufficient to break a fragile glass. But it is not to be treated in such a way as to trivialize the various conditionals; as I am using 'suitably', it is not open to me to say, merely because a glass has not

[8] This point was drawn to my attention by Mellor.

[9] This is similar to a point stressed by H. P. Grice, referred to in ch. 3.

broken when knocked, that it cannot have been knocked suitably.)

The suggestion, then, is that if 'M' is a sure-fire singly-manifested dispositional predicate (e.g. 'is fragile'), 'L' stands for what we may call the response that manifests it (e.g. 'breaks'), and 'K' for what we may call the stimulus, the conditions for its manifestation (e.g. 'is (suitably) knocked'), we have this equivalence:

$$M(x, t) = \text{If } K(x, t), L(x, t)$$

where this if-sentence is a non-material conditional. Some juggling with the time references may be needed, especially to allow for dispositions whose manifestation takes a considerable period. For example if 'M' stands for 'is radioactive, with a half-life T', and 'a' for some fairly large lump of material, '$M(a, t_1$ to $t_2)$' will represent a strictly probabilistic but approximately sure-fire disposition, analysable approximately as 'If a is left alone from t_1 to t_2, where this is a period of length t, then between t_1 and t_2 a fraction $1-(1/2^{t/T})$ of a disintegrates'.

But does this suggestion give an adequate account of the meaning of such dispositional predicates? It satisfies some basic requirements. It explains how dispositions can be ascribed to particular things at particular times and over periods, both when they are being manifested and when they are not. It allows a glass to be fragile for a while and then to be toughened by heat-treatment and to be fragile no longer. It allows a piece of wrought iron to be tough at most times but to have had a brief period of brittleness when it was cooled to the temperature of liquid air. It also allows that something may have a disposition even if neither it nor anything else ever manifests that disposition. To borrow an example from Mellor, a certain nuclear fuel may have a disposition to explode, which justifies the taking of special precautions in its use, even if (thanks to those precautions) no fuel of that kind ever does explode.

If we stick to a suppositional analysis of the non-material conditionals, this account keeps dispositional statements saying something about their ostensible subjects. By contrast, a paraphrase which Quine[10] suggests for 'x is soluble',

$$(\exists y)(Mxy \text{ and } y \text{ dissolves})$$

[10] *Word and Object*, p. 224.

where 'M' means 'are alike in molecular structure', makes dispositional statements say something about other things of the same sort. This would clearly be wrong as an analysis of dispositional statements; admittedly Quine puts it forward not as an analysis but as a proposed linguistic reform, but it is an unnecessary one. Similarly Smart[11] suggests that we can keep a material conditional interpretation for dispositional statements, avoiding trivialization by accepting (or stressing) just those conditionals which are derivable from some more fundamental theory; we could get the same result by using a metalinguistic interpretation of the conditionals used as expansions of dispositional statements. Such approaches either mix up the task of analysing dispositional statements with that of explaining either the reasons that people have for using them or what it is about the world that makes them useful, or abandon the first task and pursue one of the others instead. But there is no need to despair of understanding dispositional statements in their own right before we go on to consider how we are able to use them and to what objective realities they point.

The suggested analysis gives, I think, a meaning which dispositional statements *may* have: when this is what is meant, I shall say that what is being ascribed is a *minimal disposition.* The ascription of a minimal disposition is equivalent to the asserting of a non-material conditional or of a set of such conditionals. However, it seems that dispositional statements often assert something more than minimal dispositions. Quine[12] says that 'a theory of subvisible structure' is 'intruded' when dispositions are spoken of, 'a hidden trait of some sort . . . that inhered in the substance and accounted for' the manifestation, such as dissolving on immersion. Goodman[13] goes further. 'If certain other manifest properties are somehow connected with flexing, not merely casual accompaniments of it, exhibition of these properties by a thing not under pressure will be grounds for regarding the thing as flexible. In other words, we can define "flexible" if we can find an auxiliary manifest predicate that is suitably related to "flexes" through "causal" principles or *laws.*' Armstrong[14] says that 'to speak of an object's having

[11] 'Dispositional Properties', *Analysis*, 22 (1961–2), 44–6.
[12] Op. cit., p. 223. [13] Op. cit., p. 48.
[14] *A Materialist Theory of the Mind*, p. 86.

a dispositional property entails that the object is in some non-dispositional state or that it has some property (there exists a "categorical basis") which is responsible for the object manifesting certain behaviour in certain circumstances'; but he says also that 'we may not know anything of the nature of the non-dispositional state'. He also gives an *a priori* argument which purports to prove that we can speak about dispositions only if we are prepared to suppose that there is some such 'basis', even if we do not know its nature. Similarly Harré[15] says 'the reason why we believe that a certain disposition can be truly asserted of a thing or material is that we think or indeed know that it currently has such and such powers', while he analyses the ascription of a power thus:

> X has the power to A = if X is subject to stimuli or conditions of an appropriate kind, then X will do A, *in virtue of its intrinsic nature.*

Harré is equating a disposition with what I have called a minimal disposition, but he is saying that wherever a disposition can be asserted truly, a power can also be ascribed, and a power-ascription includes a reference to the thing's intrinsic nature, that is, to what Armstrong calls the basis of the disposition.

As I have indicated, three sorts of question may get mixed up here:

(i) When we speak about dispositions and powers, are we ascribing minimal dispositions only, or are we (as Quine and Armstrong suggest) saying also that they have some possibly unknown basis?

(ii) Must we, as Goodman's remark suggests, know of some basis in order to be able to ascribe a disposition?

(iii) Is there, whenever the ascription of a disposition or power is justified, some basis, some intrinsic nature that accounts for the thing's having this disposition?

There is a further question whether (in each case), if a basis is required, it must be 'categorical' and 'non-dispositional'.

As for question (i), although Goodman speaks about *defining* 'flexible' and about using the auxiliary manifest predicate as *definiens*, it is clear that this cannot be the meaning of 'flexible', but only another term which is coextensive with it. We can

[15] Op. cit., p. 85.

know what 'flexible' (or 'fragile' or 'soluble' or 'irritable') means even if we do not know any manifest feature which is found in all and only flexible (etc.) things, and certainly without knowing what intrinsic molecular structure would account for flexibility, what neurophysiological condition or even what psychic tensions would explain why someone is irritable. No specific basis can be even part of what a dispositional predicate means. On the other hand, it is quite plausible to maintain that it is part of the meaning of a dispositional term that *some* basis is present. That is, we might borrow Harré's phrase and define 'fragile', for example, as follows:

> 'x is fragile at t' = 'If x is suitably knocked at t, x breaks at t, in virtue of its intrinsic nature',

or, more simply,

> 'x is fragile at t' = 'x is such that if it is suitably knocked at t, it breaks at t'.

That this is a different meaning from the minimal disposition one can be shown by an example. We can surely understand the following story:

There are two glasses, *a* and *b*, both made at time t_1, which not only look alike but *are* alike in all their intrinsic features. At time t_2, each is knocked hard in the same way; *a* breaks and *b* does not. Moreover, this would have happened if they had both been similarly knocked at any time between t_1 and t_2.

This story is no doubt surprising; we may be unwilling to believe it; we may well ask anyone who tells it how he knows that *a* would have broken and *b* would not if they had been knocked earlier. But we can *understand* what we are reluctant to *believe* and would find it hard to *support*. And this story ascribes a minimal disposition of fragility to *a*, but not to *b*, from t_1 to t_2, while it denies that this disposition has any basis in occurrent, categorical, let alone manifest, features that *a* possesses but *b* lacks between t_1 and t_2.

Since these are different meanings, we must be suspicious of any *a priori* argument which purports, as Armstrong's does, to show that the ascription of a disposition entails that it has a basis. Armstrong's initial argument is epistemological: it tries to show that we could not have any good reason for ascribing minimal dispositions on their own at times when they are not being manifested. But this is not true. Suppose that glasses *a* and *b* were manufactured in different ways, each being a member of a batch of twenty similarly manufactured ones. All the *other* members of *a*'s batch have been knocked, and have broken, at various times between t_1 and t_2. All the *other* members of *b*'s batch have been similarly knocked between t_1 and t_2 and none of them have broken. Knowing this, do we not have fairly good inductive reasons for saying, just before t_2, that if *a* is knocked at t_2 it will break but that if *b* is knocked at t_2 it will not? That is, we have inductive reasons for ascribing a minimal disposition of fragility to *a*, even if there is no intrinsic difference between *a* and *b*. Of course, in doing this we should be admitting action at a (temporal) distance, supposing that the method of manufacture before t_1 made a difference to what happened when the glasses were knocked at some time after t_1, without making any difference to their internal states in the meantime. It is hard to believe this, but there are no good grounds for ruling out absolutely and *a priori* the possibility of action at a distance. This point can be strengthened if we consider dispositions whose manifestations are less destructive than that of fragility, and which the same thing can therefore manifest repeatedly. Suppose that *a* is tapped lightly, in air, repeatedly between t_1 and t_2 and every time it gives out the note middle C. Under the same test, *b* each time gives out a slightly different note, say the E above middle C. Do we not have good inductive reasons for saying that if *a* and *b* had each been tapped lightly at some time when, as it happens, neither was tapped, *a* would have given out note C and *b* would have given out note E? Of course, we shall think it is very likely that this difference correlates with some lasting difference in structure, but we cannot say *a priori* that it must do so, nor can we say that there is any incoherence in thus arguing inductively to different minimal dispositions but simultaneously denying that there is any difference in occurrent intrinsic states.

In a second version of his argument Armstrong[16] says that 'for every true contingent proposition there must be something in the world . . . that makes [it] true'. So if a conditional such as I have equated with the ascription of a minimal disposition

[16] *Analysis*, 30 (1969–70), 23–6.

holds at time *t*, there must be something in the world at *t* that makes it true; this must be something actual; that is (in my example) there must be some actual property possessed by *a* at *t* but not by *b* which makes it true that *a* would break if it were struck at *t*. I concede that this would follow if a conditional was (in the strict simple sense) true and also was not predictive, retrodictive, and so on. But as I have argued that conditionals are not true, Armstrong's argument fails. He also says, however, referring to my own earlier treatment of counterfactuals,[17] that if they are taken as condensed arguments the same conclusion follows: if we expand 'If *a* had been knocked at *t* it would have broken' into an argument we should need the premisses '(1) All pieces of glass that are of the sort *X* and are struck break; (2) This piece of glass is of the sort *X*', and here (2) would assert that the glass has some actual, categorical, property. But this is not so; 'is of the sort *X*' might refer e.g. to the method of manufacture or the previous history of the glass, not to its present state. And in any case the counterfactual could be construed suppositionally but not as a condensed argument, and its use would not then commit its user to asserting any occurrent difference between *a* and *b* at *t*.

It is true that (as Coder[18] says in a partial defence of Armstrong) to ascribe a disposition to *x* is not merely to say that *x* would exhibit behaviour *B* in circumstances *C*: if *anything* would do the same, we should not assign the disposition to *x*. 'We do not say that an object is disposed to get wet in water on the grounds that it would get wet if immersed in water. Any object would do the same.' (This example, indeed, is not quite correct. A duck doesn't get wet if immersed in water, but only if immersed in a detergent solution. But the point is clear.) To ascribe a disposition *to x* may indeed be to say that *x* would respond to the 'stimulus' in a way that other things would not. But (as Coder says) this difference need not be based on any concurrent state of the object.

We cannot, then, argue *a priori* that the ascribing of a disposition either entails or epistemologically requires the claim that it has a categorical basis. But if Armstrong's thesis were

[17] 'Counterfactuals and Causal Laws', *Analytical Philosophy* (First Series), ed. R. J. Butler, pp. 66–80.

[18] *Analysis*, 29 (1968–9), 201.

watered down to saying that wherever a disposition can be ascribed, it is a plausible empirical hypothesis that it has some basis,[19] then it could be defended. But this does not determine any answer to the question about the meaning of dispositional statements.

In fact, I think that sentences with such predicates as 'fragile', 'soluble', 'irritable', and so on have the 'is such that if . . .' sort of meaning rather than the minimal disposition one. At the very least, they hover between the two sorts of meaning, with a tendency to favour the 'such that . . .' one. But I have nothing stronger than my own linguistic intuitions to support this opinion.

§ 2. *The Ontology of Dispositions and Powers*

While I have so far been considering primarily questions of meaning, I have incidentally answered question (ii) above in the negative: in order to be justified in ascribing a disposition we do not need to know of any basis, or even that it has some basis. If 'power' is defined in Harré's way, then it follows analytically that we cannot be justified in ascribing a power unless we have grounds for believing that the thing to which it is ascribed has some 'intrinsic nature' in virtue of which it will respond appropriately to the 'stimulus' or behave in the right way in the right circumstances; but again we need not know *what* this intrinsic nature is.

I have agreed that where a disposition can be ascribed, it is a plausible empirical hypothesis that it has some ground or basis. Why this is plausible is made obvious by the extremes to which I have had to go in order to describe a situation in which an unbased minimal disposition can be ascribed. I have had to postulate that two things should behave differently in the same circumstances either with no cause at all of the difference or with remote, action-at-a-distance causes only. In the first case, *a*'s behaviour would obey one quasi-law and *b*'s another, but these would be quasi-laws only: the mention of the individuals *a* and *b* would be uneliminable. In the second case, the behaviour of both things would be law-governed, but the different reactions of *a* and *b* to the test situation would stem from no

[19] As is suggested by L. Stevenson, *Analysis*, 29 (1968–9), 197–9.

concurrent difference between their states at the time, but only from what had happened to them earlier, with no mediating chain of causal links between the remote cause and the present effect. It is the implausibility of both these stories that makes so plausible the alternative hypothesis of a concurrent basis.

All this argument shows, however, is that it is plausible, whenever we ascribe a minimal disposition, to ascribe something else as well. But *what* else? Armstrong's answer is that the basis of a disposition is itself categorical, non-dispositional: it is a set of actual occurrent features. It is generally believed that what makes the difference between a fragile glass and one that has been toughened by heat-treatment is a change in the molecular structure. Similarly, the molecules in a piece of wrought iron are moving very differently at ordinary temperatures, when it is tough, and at very low temperatures, when it is brittle. For many physical dispositions anyway, molecular structure and behaviour would seem to be the categorical basis we need. The tendency for a radioactive isotope to decay at a rate measured by its half-life is a probabilistic disposition, the basis for which would be its atomic structure.

A rival view, however, is that what we need to add to a minimal disposition is not a non-dispositional basis, but something which is still characteristically dispositional while at the same time an occurrent intrinsic state of the object. Dispositions, says Mellor, are 'real states of the thing'. On the previous view, when we speak of a power to do this or a disposition to do that we are introducing some categorical property obliquely but without describing it as it is in itself anything like adequately. But on the present view the dispositional or power style of introduction captures something that is intrinsic to and distinctive of properties of this sort: they are properties of which the description 'power to . . .' or 'disposition to . . .' is immediately appropriate or adequate, and they are something other than and presumably additional to everything that constitutes molecular or atomic structure and the like.

A third view, which might either stand on its own or be combined with this second one, is that the basis for one dispositional property may be supplied by other properties that are themselves dispositional. The heat-treated glass used for car windscreens has two important dispositional properties; it is much harder to break than similar glass before treatment, and if it does break it does not splinter but goes into granular fragments. The immediate basis of these dispositions is that a 'compressive stress' has been produced in the surface; but this stress is itself a dispositional property.

The second of these views is not often explicitly adopted; but (possibly accompanied by the third) it seems to be required by Popper and perhaps by Harré as well. For, as Mellor argues, a propensity theory of probability, such as Popper puts forward, remains distinctive and resists reduction to some rival theory of probability only if the propensities which it postulates—which are defined as a species of dispositional properties—are recognized as real physical features in their own right.[20] And if, as Popper says, all physical (and psychological) properties are dispositional, there can be no non-dispositional basis for them. Only by taking dispositions as a special sort of properties in their own right (special in the sense that there might have been properties of another sort, even if there are not) can the propensity approach 'offer a new metaphysical interpretation of physics'.

Harré's position is more doubtful. He says: 'So the verification of the ascription of a power is a two-stage process. In the first stage a test or trial shows that the specific hypothetical is true; if dynamite is detonated, there is an explosion. In the second stage the nature or constitution of the material is investigated and it is seen that it is in virtue of its chemical nature that the dynamite can explode.'[21] Now this second stage might be no more than what Armstrong would call the discovery of the specific categorical basis of the disposition. But Harré also says: 'Within this view one may see their behaviour as flowing from their natures or constitutions as consequences of what they *are*. So they must behave in the specified way, or not be the things that they are.' And throughout he stresses the term 'nature', and says that 'power talk always involves sets of non-independent predicates'. On the other hand, he says that powers are not occult qualities because they are not qualities.[22] Harré's explicit remarks do not commit him firmly to anything more than Armstrong's view; but if this is all that he intends to put

[20] This point will be more fully developed in ch. 5, § 4.
[21] Op. cit., p. 91.
[22] Op. cit., p. 90.

forward, it is hard to see why he makes such a song and dance about powers. An analysis of power talk could indeed still play a part in a realist philosophy of science; but it is very hard to see how the concept of power could 'play a central role in a *metaphysical* theory appropriate to a realist philosophy of science' (my italics) unless we go on from the first view stated above to the second.

However, the issue between these views is elusive. What exactly is the difference between treating a disposition or power or propensity as a property in its own right and saying that a (minimal) disposition has a non-dispositional basis? What is it to *be* dispositional or non-dispositional?

The most obvious contrasts are not between properties considered ontologically, as what is there, but between ways of describing them, between modes of introduction, and between kinds of knowledge that we may have about them. We might describe the difference between an iron bar at a very low temperature and the same bar at an ordinary temperature by saying—if only we knew this—how the molecules were arranged and moving. Or we might describe this difference by saying that at the low temperature the bar is such that if it is bent, it breaks. The latter is a dispositional description, the former a non-dispositional one. The latter description points to the same difference that is explicitly described by the former description, but it points to it indirectly, by way of an effect that this difference would produce if the bar were bent. Similarly, we may know the difference in question merely as the bar's being such that if it is bent, it breaks. We then know it only as a disposition whereas we know it as a non-dispositional property if we know how the molecules are arranged and move. Now it may well be that most properties are in this sense known only as dispositions, and are therefore unavoidably described and introduced in the dispositional style. But this would not make *what is there* dispositional, and it would not give the concept of disposition or power any ontological or metaphysical role.

I think the only way in which we could get an ontological contrast between dispositional and non-dispositional properties would be to say that whereas a non-dispositional basis, e.g. a molecular structure, is only contingently related to the manifestation or display of the disposition, a dispositional property is necessarily related to that display. But even this is not clear. Everyone will agree that the molecular structure of a fragile glass and its being (suitably) knocked are two causal factors the conjunction of which causes the breaking of the glass. These two factors are jointly sufficient for the breaking, but it is a contingent fact that they are so. Equally, then, the connection between the molecular structure and the fact that if the glass is knocked it breaks is contingent. Generalizing, if *A* and *B* together cause *C*, *A* is contingently related to the fact that if *B* occurs, *C* follows. We may say that *A* 'necessitates' this, but the necessity is only causal, it is a contingent fact that this causal relationship holds. But if there were a property which was in an ontological sense distinctively dispositional, it would be an *A* which would logically, not merely causally, require that if *B* occurred, *C* would follow. Just as a dispositional *description* entails the corresponding conditional, so, if any *property* were distinctively dispositional, would that property itself entail or logically require that conditional. Or, if this is dismissed out of hand on the ground that properties cannot enter into relations of entailment or logical necessity, a distinctively dispositional property would be one of which any adequate description, any account that revealed just how anything that had that property was in itself, would entail the appropriate conditional. If fragility were such a property, it would be, we may say, conditional-entailing; it would be something additional to the molecular structure (which is *not* conditional-entailing), but somehow attached to that structure.

There are at least two reasons for doubting whether there are any distinctively dispositional properties in this sense. First, it seems quite unnecessary to postulate them. Why should we insert this extra element between the non-dispositional basis and the causal behaviour? It looks like a fiction generated by treating as the name of a separate entity the dispositional style of describing the basis by way of the causal behaviour. Secondly, the suggestion that there are such properties is in open conflict with Hume's principle that there can be no logical connections between distinct existences. For if fragility in this sense were an intrinsic property of the glass, then it, the being struck, and the breaking would all be distinct existences, and yet on this view the conjunction of the first two would entail the third. The ontological claim that dispositions or powers exist as real states of things seems to result from an unhappy mixing up of two points which are separately legitimate: that a dispositional—and therefore conditional-entailing—*description* may be justified, and that a fragile glass, for example, differs intrinsically and categorically—namely in molecular structure—from a superficially similar glass that is not fragile.

Harré, indeed, suggests that the concept of 'power' has a special role to play in explanation. He says that 'a satisfactory non-Humean and realistic account of causality' involves reference to an underlying causal mechanism; but then the problem arises whether the working of this mechanism is itself merely a matter of Humean regular sequence. If it is, we have only 'pushed the irrational back a few steps but not scotched it'. And yet if we merely looked for a further, yet more underlying mechanism to explain the first we should be well launched on an infinite regress. Harré thinks that 'powers' help with this problem. 'A regress of explanation is closed, in a proper way, by adverting to entities, individuals, and materials, which are characterized solely by their powers; that is, specifically by what they can do, which we know, and only unspecifically by what they are, which for the entities which close regresses of explanation, we do not know.' But, he adds, 'Scientific knowledge also consists in the knowledge of the nature of things in virtue of which they have the powers they do.'[23]

But this is either confused or evasive. One possibility is that it is only the style of reference to these entities that is characteristically dispositional, that the entities characterized only by their powers are merely *temporarily* ultimate, they are merely the last itsems we have so far introduced in the regress of explanation. If so, this closing of the regress is only verbally satisfactory. We are just saying that there is something here which does work in the way observed or inferred. When we later discover more about the natures of these entities, the connection between their natures and what they do will still be contingent and Humean; we shall merely have taken a further step in the regress, not resolved it into something essentially different. As long as he adheres to this interpretation, Harré cannot consistently hold

[23] *The Principles of Scientific Thinking*, pp. 282–3.

that there is anything wrong with the 'irrationality' of which he complains.

Alternatively, if he postulates that there are, at some level, entities whose powers are part or the whole of their natures, then he would be introducing what I have called distinctively dispositional, conditional-entailing properties. These would indeed terminate the regress of explanation absolutely, not merely temporarily, and would give us something radically different from the 'irrational', Humean, regular sequence. And this is strongly suggested when Harré speaks about 'ultimate entities' which 'must be point centres of mutual influence, that is centres of power'[24] and indeed by the whole tone of this final chapter with its speculative claims about the ultimate constitution of the universe. But this distinctive doctrine would be achieved only at the cost of violating not only Hume's outlying theses but also the central and overwhelmingly plausible one that there are no logical connections between distinct existences. Harré seems to want to have it both ways, both to embrace the distinctive speculative doctrine of entities that really are ultimate in themselves, and to shelter in the safety of the view that their ultimacy is only relative and temporary, that their being centres of power is no more than our interim way of introducing them. It is hard to pin him down to explicit endorsement of the implausible extreme of anti-Humeanism; but unless he does endorse it he has no right to suggest either that he has a dramatically new view of the world or that he has found a way out of the supposedly objectionable regress.

We may well sympathize with Harré's desire to find something in causal processes over and above regular sequence, and certainly Hume's central and clearly defensible principle does not preclude our doing so. But powers or intrinsically dispositional properties are not the answer.

Popper's account of probability as a propensity is complicated by his insistence that this, while objective and physically real, is a 'relational' property of an experimental set-up—that is, it is a property of the whole set-up which consists in relations between its constituents. He likens it to the fact that the masses of the planets are negligible compared with the mass of the sun, which is a property of the whole solar system constituted by

[24] Op. cit., ch. 11, esp. p. 308.

relations between its parts. There can be no objection to the reality of relational properties in this sense, but it is very hard indeed to see how a propensity to generate, say, certain frequencies of heads and tails could be *constituted* by relations between a penny and the method of tossing it. It is, again, more plausible to say that there is a causal or quasi-causal fact about this set-up, that it would generate such-and-such frequencies for heads and tails, that this causal fact is explained by certain occurrent, non-dispositional features of the set-up—just how the penny is placed on the thumb, whether it is symmetrical or not, and so on; and among these certain relational features are important, such as the relation between the mass-distribution of the penny and the thrust of the thumb—but that we should be mixing up the causal fact and the occurrent features if we postulated, as a present relational property of the set-up, a propensity to generate these frequencies.

It might be objected that in setting up the alternatives that a property either is what I have called conditional-entailing or is only contingently related to the relevant conditional(s) I am neglecting a vital third possibility. Could not a property be dispositional in the sense that while it could not be said to *entail* the relevant conditional(s), it could still necessitate them with something stronger than a causal relationship of a Humean sort? If the conjunction of *A* and *B* would lead to *C* by some kind of natural necessity, there would also be a sort of natural necessity in the link between *A* and the conditional 'If *B*, then *C*'. We might, then, try to define a distinctively dispositional property as one such that the conjunction of it with the 'stimulus' would naturally necessitate the 'response'. Such a property would be conditional-necessitating though not conditional-entailing, and the criticisms I have directed against the latter possibility would not affect it.

Now if there are relations of natural necessity, presumably all causation includes them; then, since presumably all properties can enter (as causal factors) into causal processes, all properties would indeed turn out to be dispositional in this sense. Whether there is something appropriately called natural necessity, which a Humean or regularity account of causation leaves out, is an important question, in a way the central question for a study of causation. I cannot discuss it here. If it were answered in the affirmative, it would indeed give us an interesting respect in which (all) properties could be called dispositional. But this would mean, for example, that the molecular structure of a fragile glass was *itself* a dispositional property. This view would not require or enable us to postulate fragility as an additional dispositional property over and above the molecular structure. So, while I cannot now rule out this possibility, I do not think the fulfilment of it would provide what is wanted by those who would treat dispositions themselves as real states of things, additional to molecular structures and the like.

What about the suggestion that the basis of one dispositional property might be other *dispositional* properties? Since I have rejected the view that dispositional properties form any special ontological category, this could now mean only that the basis of one disposition was described in a dispositional style but with reference to other sorts of manifestation. For example, the toughness of a windscreen has as its basis compressive stresses in the surface of the glass; any such stress is the disposition of small portions of the surface to expand sideways if the neighbouring portions are removed. Since such stress can be measured optically, by the amount of double refraction, we could also introduce the basis of the toughness disposition as the glass's being such that it doubly refracts light in such-and-such a way. For many of our purposes, theoretical as well as practical, we can move around in this sort of manner from one dispositional description to another. Yet since no properties are dispositional in themselves, these dispositional descriptions reflect the presence of features which in themselves are categorical. What is there must be distinct from all the things that it would do.

When we thus insist on an occurrent or categorical basis for dispositions, we cannot require that this basis should consist of *manifest* properties. Many dispositions were known before their categorical bases were discovered; for others the bases are unknown to us now and may always remain so. The ontological case for an occurrent basis is to be sharply distinguished from a semantic argument that the meaning of an explicitly dispositional predicate must be derived from that of one or more manifest predicates. Since to call something flexible is to say that it would bend, or that it is such that it would bend, in certain circumstances, the meaning of 'flexible' is derived from the meaning of 'bend'. Of course, we can use already-derived dispositional predicates to derive further, second-order dispositional ones. Thus 'magnetic' is a dispositional predicate, 'this is magnetizable' means 'this (is such that it) would become magnetic if placed in a (sufficiently strong) magnetic field': 'magnetizable' then, is a second-order dispositional predicate.[25] And in principle this process could go on to any number of further stages. But the meaning of a dispositional predicate of whatever order must be derived ultimately from that of the manifest predicates with which the series starts. This is correct, but it is quite a different matter from the plausibility of the empirical hypothesis of a categorical basis for each minimal disposition. Dispositions need to be anchored in two quite different ways: semantically, to manifest predicates from whose meaning that of dispositional terms is derived, and ontologically, to a basis that is usually not manifest and is often unknown, but which causally accounts for the behaviour by which the disposition is manifested.

To sum up, there are three possible ontological views about dispositions. The first is the one Armstrong calls phenomenalist and ascribes to Ryle: we attribute a minimal disposition, which is in effect to assert a conditional or set of conditionals, themselves to be interpreted as inference-tickets; but this does not mean that anything is going on in the thing to which we attribute the disposition which is not going on in similar things from which we withhold this description. The second is the 'realist' view, that dispositions have occurrent (and concurrent) categorical bases consisting of properties which are not in themselves peculiarly dispositional, though they may be introduced in the dispositional style and may be known only as the bases of these dispositions; although the dispositional descriptions are conditional-entailing, the properties to which they point are only contingently related to the displays of the dispositions. The third is what we may call the rationalist view; dispositions (while still being intrinsically dispositional and conditional-entailing) are real occurrent states of the object, different from anything that the realist would call a categorical basis (which may or may not be there as well), but actually

[25] Cf. Broad, *The Mind and Its Place in Nature*, p. 432.

present both when the disposition is being manifested and when it is not. Of these, the third view can be firmly rejected. The second cannot be shown to be required *a priori* for every ascription of a disposition, but wherever such an ascription is in order, it will be prima facie a plausible hypothesis that it has the sort of basis suggested by the realist view. The first view may sometimes be right: a bare minimal disposition description might be in order without there being any occurrent (or at any rate concurrent) basis. But this is in general unlikely, and we certainly cannot infer from the fact that a description is dispositional that it is wrong to expect that it has a categorical basis.

We can now see what is, and what is not, wrong with dormitive virtues and suchlike occult qualities. Anyone who has discovered that opium puts people to sleep has a good reason for making the further contingent assertion that it has a dormitive virtue in the sense that it has some as yet unknown—but not unknowable—property or constituent which is causally responsible for this effect. There is nothing absurd or improper in postulating this, and (as Harré says) this is just how scientists do proceed; such a postulation leads on to further discoveries. It is indeed a mistake to offer this dormitive virtue, postulated simply because opium has been found to put people to sleep, as an explanation of why it does so. *As an explanation* this is empty; it is only a place-holder for a genuine explanation. Yet it points the way towards a genuine explanation, whereas the simple statement that the taking of opium is regularly followed by sleep does not. But a quite different point is this: if someone postulates a dormitive virtue in accordance with what I have called the rationalist view of dispositions, that is as something whose intrinsic nature, adequately described, would *entail* that if anyone consumes a fair amount of it he falls asleep, then this is wrong in another way. There are no such intrinsically dispositional properties as this is supposed to be. The explanation given by postulating this one would not be empty, but too good. It does not merely re-state what is to be explained, it is not merely the promise of a detailed explanation that has still to be discovered; it is an all-too-perfect explanation which usurps the place of any merely contingent one. What would be the point of showing that opium contains morphine which (despite its name) is only contingently related to sleep, if we knew already that opium contained an intrinsic power whose presence entailed the production of sleep?

I have so far been concerned mainly with sure-fire dispositions; but much the same can be said about probabilistic ones. 'He is irritable', construed as a minimal dispositional description, will now mean 'If he is disturbed or provoked he is likely to become angry'. Construed non-minimally, it will mean this with the tag 'in virtue of some present intrinsic nature or state', or, more briefly, 'He is (now) such that if he is disturbed or provoked he is likely to become angry'. As before, the dispositional statement seems to hover between these two meanings, but with a bias towards the non-minimal one. With probabilistic dispositions too it is plausible to propound the realist hypothesis that they have concurrent bases related contingently to their manifestations. The contingent relation might be of either of two kinds. The object's (or person's) behaviour might be indeterministic, so that the conjunction of the basis with the 'stimulus' would not causally ensure the 'response': the 'response' would still be subject to some element of chance or free will. Or it might be that the antecedent of the conditional (e.g. 'he is disturbed or provoked') covers a range of 'stimuli', some of which conjoined with the basis would causally ensure the 'response', while others would not; but the ascriber of the disposition is not able to classify these stimuli separately: he cannot sort out the kinds of disturbance, say, that will ensure that this man becomes angry from those that will leave him calm. (These are two different ways of ascribing a probabilistic disposition; a distinct possibility is that a speaker should tentatively ascribe a sure-fire disposition, and the same words 'If he is disturbed or provoked he is likely to become angry' could bear this meaning too.)

The rationalist view which I have criticized could be taken also of probabilistic dispositions. But now it would mean that the disposition (propensity) was an intrinsic state of the object the conjunction of which with the 'stimulus' would only probabilify or quasi-entail the 'response'. But my objection to this sort of view still stands. The conjunction of this intrinsic state with the 'stimulus' would still be a distinct existence from the 'response', and if probabilification or quasi-entailment is a *logical* relationship, it could not hold between them any more than entailment could. But if 'probabilification' is taken in a weak sense, as a contingent relation, then indeed this conjunction could probabilify the response; but we should now have gone back from the rationalist to the realist view.[26]

We may now relax also the restriction to singly-manifested dispositions. As I said, the distinction is somewhat arbitrary. Is (inertial) mass a singly- or a multiply-manifested disposition? It is displayed both in the difficulty of setting the massive object in motion, if it is at rest, and in the difficulty of stopping it or turning it aside if it is moving. The statement that *a* has mass *m* entails an infinite set of non-material conditionals of the form 'If *a* is at rest/moving in such-and-such a way and such-and-such a force is applied to it such-and-such happens'. But of course all this can be summed up in a single Newtonian formula 'If a force *f* is applied to *a* (and no other forces are), *a* will have an acceleration measured, in the right units, as *f*/*m*'. If belief is a disposition, it is a multiply-manifested one. *A*'s belief that it will be fine tomorrow may be displayed by his saying that it will be (to his friends) or that it will not be (to his enemies), by his leaving his only raincoat at the cleaners, buying a ticket for an excursion, adjusting his lawnmower, singing happily to himself, cursing under his breath, and so on. It may also be displayed by his acquiring another disposition, surprise, when he wakes up and sees the rain pouring down. Many of these manifestations can be summed up in the formula: *A* acts in ways which, if and only if it is fine tomorrow, will help to satisfy some desires he now has. But some of the manifestations listed escape this formula—the singing, the cursing, the surprise. Now it is in no way surprising that some one state should have (when combined with various 'stimuli') a number of different effects (or probabilistic quasi-effects), whether or not these can be summed up under one formula. So it is not surprising that a property can be introduced in a dispositional style by reference to various different manifestations. A thing or a person may well be now such that if *P*, *Q*, but also, in virtue of the *same* intrinsic condition, such that if *R*, *S*, and so on. But a multiply-manifested disposition would be surprising if we took the dispositional description to be immediately revelatory of the

[26] The strong and weak senses of 'probabilification' will be examined in ch. 5, § 2, pp. 171–2.

intrinsic condition itself. It would be hard to see how one and the same belief-disposition could be essentially a tendency to make predictions (of two opposite sorts), to buy tickets, to sing, and so on. This is a difficulty for the 'rationalist' view of dispositions which I have already rejected on other grounds. For the realist ontology, multiply-manifested dispositions are no problem. But for the phenomenalist view, a multiply-manifested disposition should be a contradiction in terms. If a disposition consists just in the holding of a conditional, there should be just as many dispositions as there are conditionals, that is, as there are manifestations. It is very strange to say, as Ryle[27] does, that 'to believe that the ice is dangerously thin is to be unhesitant in telling oneself and others that it is thin . . . But it is also to be prone to skate warily, to shudder, to dwell in imagination on possible disasters . . . It is a propensity not only to make certain theoretical moves but also to make certain executive and imaginative moves, as well as to have certain feelings.' As minimal propensities, these are all different: if there is nothing but the minimal propensities there, how can they all be one propensity? It is not enough to say that 'these things hang together on a common propositional hook. The phrase "thin ice" would occur in the descriptions alike of the shudders, the warnings, the wary skatings. . . .' This phrase might or might not occur in these, but it also occurs in descriptions of things that don't constitute this belief, e.g. of someone wishing that there were some thin ice about. A more important way in which the thin-ice-belief propensities hang together might be that they show strong (but not perfect) positive correlations with one another, that those who skate warily also imagine disasters, and so on. But even this does not explain why on the phenomenalist view we should think of a single disposition, multiply-manifested, rather than of a cluster of different dispositions.

But the realist view can explain this, and it can also explain how if people adopt that view they can form a single dispositional *concept* related to a multiplicity of manifestations. Temperature is a good example. Samples of many materials, after being placed near a fire for a short time, feel warm to the touch; left there longer, they feel hot. Simple correlations can be set up: the hotter things feel, the more rapidly they melt ice

[27] *The Concept of Mind*, pp. 134–5.

or snow, the more they tend to scorch things that touch them, the more they make objects in contact with them also feel hot, and so on. A pre-scientific scalar concept, of things being more or less hot, is already one of things being in this or that state upon a scale of states, where each state on the scale has a number of correlated effects. Elementary science adds to these correlations: many materials expand when hot, and expand the more the hotter they are. Thus arises an elementary scientific concept of temperature as of a scale of states which are known primarily as dispositions to produce a variety of correlated manifestations, but also as typical products of larger or smaller doses of certain similar treatments, such as placing the objects near a fire, in strong sunlight, or in contact with other objects at a high temperature. At this elementary stage of science all that is known about temperature is the network of correlations; but these correlations are good enough to support the hypothesis that there is some single scale of internal states which both are produced by the various processes we call heating and produce the various manifestations of heat. At this level of knowledge we have a mainly dispositional concept of temperature; the disposition is multiply-manifested, but the concept is of a single scale of as yet unknown states. (I say *mainly* dispositional to allow for the fact that temperature is known not only as a cause of certain effects but also as an effect of certain standard causes.)

While we can explain the meaning of the terms 'temperature', 'hot', and their various further specifications along these lines it is highly questionable whether 'believes that' should be explained in an analogous way. There is no harm in saying that to believe that . . . is to be in such a state that . . . but it is more doubtful whether this is what 'believes that . . .' *means*. To discuss what it does mean would take us too far afield, but the basic point is this: despite all the criticisms made in recent philosophy of the notions of introspection, private languages, and the like, it is a hard fact that we know more about the state of believing, as it is in itself, than we do about the states that form the temperature-scale as they are in themselves, and when we speak about believing we are using (among other things) this more intimate knowledge.

Just as, on the realist view, there is nothing surprising about a multiply-manifested disposition, so also it will not be surprising if a disposition has more than one ground or basis. A plurality of causes is as normal as a plurality of effects (in different circumstances) of the same partial cause. The fragility of glass may be based on one sort of molecular structure, the fragility of pottery on another. Even in the same material, the same disposition may have more than one ground. A piece of cloth may absorb water in two ways, by the water being taken into the individual fibres and by its being held in spaces between the fibres: its absorbency then has two different bases, the molecular structure of the fibres and the larger-scale structure in which those fibres are spun and woven. And someone's irritability may well be (immediately) based on a number of different psychic tensions, worries, or frustrations.

§ 3. *Are Physical Properties Dispositional?*

My thesis that no properties are in themselves dispositional is in apparent conflict with Popper's thesis that all physical properties are dispositional—but are relational dispositions. Let us see whether we can settle or resolve this conflict in the case of, say, inertial mass.

Within Newtonian mechanics, to ascribe a certain mass to an object is at least, and primarily, to say that an infinite set of conditionals holds, each of which has the form 'If such-and-such a force (alone) is applied to this object, such-and-such an acceleration results'. Elsewhere in the theory there will be ways of fixing the meaning of 'acceleration' and—what is more difficult—of specifying when such-and-such a force is applied. But the meaning of the term 'mass' is tied up with the 'law' which connects it with force and acceleration.

Is the ascription of mass only the ascription of such a minimal disposition? This depends on the philosophy of science explicitly or implicitly adopted by the ascriber. Someone who takes a positivist view will say that the holding of the appropriate conditionals (with, indeed, further unpacking of the terms 'force' and 'acceleration' in them) is all that can be checked, this is all that can be meaningfully asserted, and certainly all that he himself would assert. Someone who takes a realist view will suppose that an object's mass is some permanent occurrent feature of the object, so that in ascribing a mass he will be saying that the object is *such that* these conditionals hold, but (consistently with what we have said about the realist view of the grounds or bases of other dispositions) that whatever this phrase 'such that' points to is only contingently related to the force-acceleration conditionals. A variant of the realist view would be that this phrase 'such that' points only to the presence of other *dispositional* properties, that we can link the mass of an object with the sum of the masses of the parts into which it could be divided, and so eventually of the atoms composing it, the mass of each atom can be related to its sub-atomic structure, and so on. Mass is, we might say, a well-behaved property, there are systematic regularities connecting its various specific instantiations with one another and with other dispositional properties, but what we deal with and what we find regularities within is only a network of dispositions, we never get to grips with a non-dispositional basis of any of these. We cannot call them *minimal* dispositions precisely because they are connected by these regularities: each of them is appropriately ascribed by a 'such that if . . .' formula; but when we trace the implications of the 'such that' in each case all we come to are further dispositional ascriptions. Someone who takes what I have called a rationalist view will treat mass as a property which an object has in itself, which is inevitably a distinct existence from most of the force-acceleration combinations which would reveal it (since at any one time the object can be subject to only one (possibly zero) resultant total force and can have only one (possibly zero) acceleration), and yet whose presence entails all the conditionals connecting resultant force with acceleration.

As I have said, what an ascription of mass means depends on the ascriber's philosophy of science. The interesting question is whether one of these views is more defensible than the rest. The reasons already given for rejecting the rationalist sort of view still seem overwhelming; but we have a serious choice between the others.

It is worth stressing here the empirical facts which even a minimal dispositional account of mass presupposes; these are very well brought out by Nagel,[28] for example.

First, it is an experimental fact (or an idealization of one) that if two bodies *A* and *B* are affected only by each other, they set up accelerations in each other opposite in direction and that

[28] *The Structure of Science*, pp. 193–6.

the ratio k_{BA} between the acceleration of A and the acceleration of B is invariant, whatever their positions and velocities and modes of mutual influence may be. Similarly if B is taken along with any third body C there is an invariant ratio k_{CB} between their mutually induced accelerations. It is a *further* experimental fact (or idealization of one) that if A is then taken along with C, the invariant ratio k_{CA} between the mutually-induced accelerations of A and C will equal the product of k_{CB} and k_{BA}. And it is yet a *further* experimental fact (or idealization of one) that if B and C are combined into one body $B+C$, and this is taken along with A, the invariant ratio $k_{(B+C)A}$ between the mutually-induced accelerations of A and of $B+C$ will equal the sum of k_{BA} and k_{CA}.

It is upon all three of these facts that the quantitative definition of inertial mass depends. Given that they hold, we can define the relative mass (or mass-ratio) of B to A as k_{BA}, of C to B as k_{CB}, of C to A as k_{CA}, and of $B+C$ to A as $k_{(B+C)A}$, with the assurance that these relations of mass will form a constant and coherent system: the mass-ratio of C to A is the product of those of C to B and of B to A, the mass-ratio of $B+C$ to A is the sum of those of B to A and of C to A. Mass-ratios thus defined, in other words, behave as *ratios* should. We can then choose any object, say A, as fixing the unit of mass, and then measure the mass of B by the mass-ratio of B to A, and so on.

It is because all three empirical facts hold for all such pairings of material objects that the concept of inertial mass, defined even dispositionally, just by sets of conditionals, is a conveniently workable one. This clarification of the definition of mass within Newtonian mechanics is due essentially to the arch-positivist, Mach, and a positivist would say that it is just this network of empirical regularities that tempts us into thinking of a certain mass as an intrinsic property of each object. But a realist could agree with all of this, merely adding that this is a case where it is virtuous to succumb to temptation. The postulation that each material object has some intrinsic quantitative feature which reacts contingently but lawfully with imposed forces is just what we need to explain the otherwise remarkable coincidence that the three empirical facts stated above hold for all objects. The mass-ratio, e.g. k_{BA}, as defined above, is invariant just because it results from invariant but otherwise unknown lasting features that A and B separately have. It is because the mass-ratios k_{BA}, k_{CB}, and k_{CA} all arise thus from intrinsic features that they are well-behaved in the respect that $k_{CA} = k_{CB} \times k_{BA}$. And it is because these intrinsic quantitative features of B and of C are (except in some very special cases with sub-atomic particles) additive that the third empirical fact about mass-ratios holds, that $k_{(B+C)A} = k_{BA} + k_{CA}$.

The old definition of mass as quantity of matter is, then, not so far from the mark after all. It is true that we do not know much about what matter is; but it is reasonable to postulate that there is a relatively permanent quantitative something-or-other intrinsic to objects and additive in all their normal combinations. In saying that an object has such-and-such a mass we may reasonably opt for the interpretation that this is to say that it is such that a certain set of conditionals holds, and that although this style of introduction is dispositional, what is introduced is an intrinsic, quantitative, but otherwise mainly unknown feature.

It is not completely unknown: the mass of certain particles is due, we say, wholly or partly to their electric charge. But of course to say this is only to link the mass disposition with another set of dispositions. We do not know of any non-dispositional basis of mass analogous to the sort of molecular structure that seems to be the non-dispositional basis of fragility. But just as it would have been reasonable to postulate that fragility had *some* non-dispositional basis before we knew *what* it was, so it is now reasonable to postulate some as yet unknown non-dispositional basis of mass.

We do not for a moment suppose that the mass of any large-scale object is an unbased minimal disposition. That is, we do not suppose that two objects can be alike in all their present intrinsic features but have different masses, so that they will react differently when pushed in the same way. But the immediate basis of the mass of a large object is the masses of its small-scale components. The crucial question is whether when we get down to sub-atomic particles it still is reasonable to suppose that the unequal mass-ratio of a proton and an electron has a basis in something intrinsic to each particle separately. It seems to me that it is. The alternative, positivist, position is to say that the experimental fact that such particles respond differently to equal forces, and the inductive projection of this into the ascription of minimal dispositions, that they would react in such-and-such different ways if various forces were applied to them, must be taken as a brute relational fact with no foundation in what each particle is in itself. And this seems utterly implausible.

What I have called a variant of the realist view—accepting a 'such that . . .' interpretation but taking this phrase to point only to other still dispositional properties—does not seem to be a satisfactory alternative. If these properties were taken as dispositions in the rationalist sense, all the difficulties of that view would be incurred. If, alternatively, they are taken as dispositions which avoid being minimal dispositions only by their mutual connections—each of them can be introduced by a 'such that . . .' phrase which points to some collection of the others—then the whole network of regularities is left unexplained whereas the postulation of an occurrent, non-dispositional basis for, e.g., mass provides at least the sketch of an explanation.

Some elaboration of this account will indeed be needed. The realist view so far stated postulates a basis for the rest-mass of an object. The increase of mass with velocity yields an additional basis. A rapidly moving object's being such that the appropriate conditionals hold consists partly in its having a certain 'quantity of matter', partly just in its moving with that velocity; the mass as a whole, of course, is relative to a certain co-ordinate system, because the velocity is so, but the rest-mass is invariant between co-ordinate systems and its basis, 'quantity of matter', can be taken as an absolute and intrinsic feature of objects. It is, of course, no more surprising that mass should have more than one sort of basis, 'quantity of matter' and velocity and perhaps more besides, than that fragility or absorbency, for example, should; we have already noted that the fragility of glass may have a different basis in molecular structure from the fragility of pottery, and that the absorbency of some woven material may be due partly to water being held in the spaces between fibres and partly to its soaking into the individual fibres themselves.

What I have said about everyday dispositions such as fragility does not, then, need revision when we come to such centrally important physical dispositional properties as mass.

We can resolve the apparent conflict between the claims that all physical properties are dispositional and that no properties are dispositional by conceding that most physical properties, including mass, are introduced in a dispositional style and are known only or mainly by way of the effects they have or would have in such-and-such circumstances, but insisting that it is reasonable to postulate bases for them which are not distinctively dispositional, which are neither mere networks of minimal dispositions nor properties whose presence would entail the conditionals by which the dispositions are displayed.

5

CONCEPTS OF PROBABILITY

§ 1. *The Multiplicity of Probability Concepts*

THE concept of probability is an unusually slippery and puzzling one. For one thing, it seems to hover uncertainly between objectivity and subjectivity. Talk about something being probable or likely seems to reflect some mixture of knowledge and ignorance—if there were an omniscient God, it is hard to imagine that he would regard anything as merely probable—and yet most of our probability statements seem to claim some objective or at least inter-personal validity, and we treat probabilities in many cases as being measurable and calculable in a strict mathematical way. In this uncertainty of status probability resembles moral concepts such as goodness and obligation, and indeed there are systematic parallels to be found between rival ethical theories and rival theories of probability.

Traditionally, adherents of each of these rival theories have tried to show that their own is the one correct account of probability and that the others are wrong. More recently, it has become fairly common for philosophers to suggest that we should find room for more than one concept of probability: Carnap, for example, held that there are just two distinct concepts of probability, one being that of degree of confirmation, the other that of relative frequency in the long run, though there have been more than two ways of trying to explicate these concepts.[1]

But two, I think, are not enough. There are at least five fairly clearly distinguishable probability concepts, with subdivisions within several of them, as well as links between them.

Faced with such a multiplicity of concepts, we may well want to examine their merits and demerits; but these may be of at least three kinds, which we should keep separate as far as we

[1] R. Carnap, *Logical Foundations of Probability*, ch. 2.

can. First, there may be questions about the internal health of a concept, whether it is coherent, and whether it is sufficiently clear and precise. Secondly, once a concept has been proposed and clarified and developed, there is the question whether it is what we ordinarily mean by a certain word or set of words—that is, whether it is *the* concept, or *a* concept, that we actually use in a certain field of thought. Thirdly—and this question can properly be raised whatever we have said in answer to the second question—there is the question whether a certain concept has a correct or legitimate application, whether there is anything in the world—as contrasted, perhaps, with our established ways of thinking and speaking about the world—which conforms to that concept, or which we have reason to believe to conform to it.

I shall argue that none of the five main probability concepts (or, with one or two exceptions, their sub-concepts) can be ruled out on grounds of internal health: they can all be coherently developed and made at least reasonably clear and precise. I shall also argue that no one (or even two) concepts cover even all the 'ordinary' uses of probability language, let alone some of the more sophisticated ones introduced by mathematicians and philosophers; in fact room will have to be found within 'what we mean by probability' for at least five basic senses and perhaps more. On the other hand, I shall argue that this is not a sheer multiplicity of radically different meanings: 'probable' and its associates are not simply ambiguous. We can understand their complex of meanings as growing out from and being nourished by a central root of meaning: they can be explained as the result of a series of natural shifts and extensions. Probability illustrates the thesis of J. S. Mill (recently taken over by Wittgenstein and his followers in their talk about family resemblances) that 'names creep on from subject to subject, until all traces of a common meaning sometimes disappear'.[2]

But precisely because this is so, the risk of falling into fallacies of equivocation, and the need to check that arguments about probability really go through with the intended interpretations of the probability statements they involve, are particularly

[2] J. S. Mill, *System of Logic*, Bk. I, ch. 1, § 5. Cf. L. Wittgenstein, *Philosophical Investigations* I, 66–7.

acute. The task is not merely to make sure that one does not slide from one sense of 'probable' or 'probability' to another. Some such transitions are valid or defensible, and the tricky problem is to decide whether a particular transition is one of the justifiable ones or not. I shall spend some time examining from this point of view some particularly important forms of probability argument, such as those that make use of Bernoulli's Central Limit Theorem and Bayes's First and Second Theorems. But I shall do this with the very minimum of mathematics. It seems to me that while there are many books that deal well with the mathematics of probability, none deal adequately with the philosophical problem of finding one's way through this maze of concepts; but for solving this problem very little mathematical calculation is needed, and any more than the bare minimum tends only to conceal the philosophical issues.

To the question whether each of the main probability concepts has correct applications I shall give a more tentative reply, but I shall at least indicate some of the considerations that favour one answer or the other.

The five concepts of which I am speaking are these. First, there is what seems to be the most natural everyday sense of such words as 'probable' and 'likely', the sense which is explained and emphasized by the 'informal' approach to probability, and of which at least a rough indication is given if we say that 'Probably *P*' or 'It is likely that *P*' means something like 'It is reasonable to believe that *P*, but the reasons are inconclusive'. Secondly, there is the 'range' or 'measure' concept: a primitive variant of this is the classical definition of the probability of an event as the ratio of the number of favourable possibilities—that is, the ones whose fulfilment would satisfy the event-description in question—to the total number of possibilities, that is, of the basic alternative outcomes. Thirdly, there is the concept of a logical relation, a quasi-entailment between data or evidence and a conclusion or hypothesis: Carnap's probability_1 or degree of confirmation. Some subdivision of this concept may be required: there are, prima facie at least, two different kinds of non-deductive support.[3] Fourthly, there is the concept of frequency, with which probability has often been

[3] J. L. Mackie, 'The Relevance Criterion of Confirmation', *British Journal for the Philosophy of Science*, 20 (1969), 27–40, esp. 37–8.

identified. Here too subdivision is needed. There is the straightforward concept of a finite class ratio, of the proportion, among the members of the class of *A*s, of those that have the characteristic *B*. There is also the more elusive concept of a limiting frequency in an infinite or at least indefinite series or 'collective': Carnap's probability_2 or 'relative frequency in the long run'. Fifthly, there is the concept of objective chance, of something which (unlike frequency) belongs to each individual event or 'trial' in itself, apart both from anyone's limited knowledge and from any class or sequence to which the individual events belong. This notion too seems at first to admit of subdivision, but I shall argue that only one form of it remains distinct from all the other concepts mentioned.

If one were concerned merely to multiply concepts it would be easy to add two more. There is the axiomatic concept, which identifies probability either with a purely formal item, the uninterpreted subject of a purely mathematical calculus of probability, or, liberally, with anything whatever which turns out to be a valid interpretation of this formal item, that is, anything at all to which the calculus can be successfully applied. Also, there is the subjective concept, developed by proponents of the subjective theory.[4] Such a probability represents the degree of belief which someone has in some matter, and, it is suggested, can be measured by the betting quotient, the rate at which this person would be willing to accept bets for or against the possible outcome in question, subject to the constraint that his betting behaviour should be rational in that it should not allow anyone to place a set of bets with the original gambler in such a way as to be sure to make a profit whatever happens, to 'make a Dutch book' against him. I leave the first of these aside because (although it is an established use of probability terminology in mathematical contexts, and is a particular specimen of a well-known kind of meaning shift by which terms are transferred from concrete interpretations to formal items in a bit of abstract mathematics and then back to *any* satisfactory interpretations of those items) it is merely an

[4] e.g. F. P. Ramsey, B. de Finetti, L. J. Savage; examples of their writings are collected in *Studies in Subjective Probability*, ed. H. E. Kyburg Jr. and H. E. Smokler. It seems to me that de Finetti comes closer than other writers of this school to what I call pure subjectivism, suggesting that even the 'logical laws' which characterize coherent opinions have only a psychological basis.

evasion of the philosophical problems about competing senses of 'probability'. I leave the second aside for a different reason. A *pure* subjectivist meaning of 'probable' and its associates, according to which 'It is probable that *P*' was merely either a report or an expression of some degree of subjective belief, would be, as I shall show, an extreme limit which the everyday meaning of 'probable' approaches, but does not, I think, ever normally reach. It is something that we *might* have to reintroduce in constructing a revised, philosophically defensible, account of probability—as I shall also try to show (in § 9). But the concept sketched above is not a pure subjectivist one, since it is subject to the constraint of rationality in the sense indicated: in terms of it, probability is not an expression of any old subjective belief, but of the beliefs of our idealized gambler, of beliefs that are at least in a minimal respect rationally justifiable. Now this is at least a first step towards the concept of a rationally justified degree of belief, which is something that will emerge from our examination of the first of our five concepts. The sophisticated variety of 'subjectivism' is not thoroughly subjectivist in spirit, but is a minimal, semi-sceptical doctrine about how far beliefs are rationally justified: what it deals with is not, therefore, a separate basic probability concept.

I turn now to a somewhat fuller analysis of the five concepts mentioned, and an examination of what I have called their internal health.

§ 2. *Informal, Simple, and Relational Probability, and the Range Concept*[5]

The 'informal' concept suffers from some indeterminacy or internal tension. At one extreme it amounts to this: to say that it is probable that *P* is to say guardedly that *P*. This is analogous, but with the important variation introduced by the word 'guardedly', to a reassertion account of the meaning of 'true'. At this extreme the informal concept verges upon what I called pure subjectivism, for to say guardedly that *P* is just to *express*—not to *report*—a belief that *P* which falls short of complete

[5] The 'informal' approach is expounded in S. E. Toulmin, *The Uses of Argument*, ch. 2; the relational account in J. M. Keynes, *A Treatise on Probability*, esp. chs. 1–4, and in R. Carnap, *Logical Formulations of Probability*; the 'classical' account in Laplace, *A Philosophical Essay on Probabilities*, trans. E. T. Bell.

subjective confidence. At the other extreme the informal concept amounts to this: to say that it is probable that *P* is to say that there are good, but not conclusive, reasons for believing that *P*. At this end of its range, the informal concept is primarily concerned with a claim of an objective partial justification for the assertion that *P*. There are various intermediate stages between these two extremes, e.g. saying guardedly that *P* while hinting that there are some reasons to back this up. So much of the ordinary thought that is naturally expressed in probability language exhibits this hovering between the two extremes that I think it is more natural to speak of one concept which is itself in a state of tension than to speak of people having two concepts, one of guarded assertion, the other of there being good but not conclusive reasons for an assertion, though from a sharply analytic point of view the latter might be at least as correct. To show that the most ordinary use of 'probable' does carry a meaning lying between these extremes, we may contrast 'It is probable that *P*' with expressions which would more obviously count as merely saying guardedly that *P*, e.g. '*P* but don't blame me if it turns out otherwise', '*P* but I won't go to the stake for it', 'In my opinion, *P*'. Linguistic intuition suggests to me, at any rate, that 'It is probable that *P*' naturally makes more of an objective claim about supporting reasons than does any of these. On the other hand, if we contrast 'It is probable that *P*' with the explicit 'There are good but not conclusive reasons for believing that *P*', it seems to me that the former reflects more concentration on *P* itself, and less on the support for it, than the latter.

At both extremes this informal concept leaves room for one thing to be said to be more or less probable than another. A guarded assertion may be more or less guarded; one set of good but inconclusive reasons may be better or more nearly conclusive than another. Presumably, therefore, this concept also leaves room for saying that two things are equally probable, in other words for being no more guarded about one than another, or for saying that the reasons that support *P* are as good but as inconclusive as those that support *Q*. It is most obvious that there is room for this where *P* and *Q* are alternatives, rival answers to the same question. But beyond this point the informal concept does not in itself leave room for the measurement or the calculation of probabilities, or for anything but the roughest sort of comparison between the probabilities of P and R if these are *not* alternatives, but answers to different questions. To say that Rock Roi is as likely to win the Cup as the Copenhagen interpretation of Quantum Theory is to be correct, or that the probability of either is 0·57 is, on the face of it, unintelligible from this point of view. It is equally unintelligible from this point of view to say that we must add the probability measures of two incompatible alternatives to find the probability measure of their disjunction. But though the informal concept does not in itself provide for such procedures, we shall see that a metric and a calculus can be grafted on to this concept without any radical conflict. It is the good reasons end of the informal concept that allows this, and that will be of the greatest importance in my later discussion: I shall use the phrases 'simply probable', 'simple probability', and the like to refer to uses of this end of the informal concept.

The 'classical' definition of the probability of an event as the ratio of the number of favourable cases to the total number of cases is, as it stands, incomplete and indeterminate. What cases? If a card is drawn at random from a standard pack, then in one respect there are two 'cases', that it should be a heart and that it should not; only the first of these is 'favourable' to its being a heart, so the ratio, and therefore the probability of its being a heart, is 1:2. In another respect there are fifty-two cases, that the card should be the ace of spades, that it should be the two of spades, and so on all through the pack; of these cases thirteen are 'favourable' so the ratio is 13:52 or 1:4. Clearly we want to prefer the second interpretation to the first, but to do so we must qualify the bare 'ratio of cases' (or of 'possibilities') definition. We might do this in either of two ways. First, we might require that the possibilities among which the ratio is to be determined should be basic alternatives or that they should be equally specific. It is clear that the drawings of the ace of spades, of the two of spades, and so on form in some sense a set of basic alternatives, whereas drawing a heart and drawing a non-heart do not, and again that the latter two possibilities are not equally specific. (It may, indeed, be argued that what we judge to be basic or equally specific alternatives depends on our conceptual framework. If we could distinguish hearts from non-hearts, but not spades, diamonds, and clubs from one another, and were unable to count cards, then indeed we might regard 'heart' and 'non-heart' as basic alternatives. But any errors that resulted could properly be ascribed to the defects of this conceptual framework.) Secondly, we might require that the possibilities among which the ratio is to be determined should be equally probable (or 'equiprobable'). Now it is not difficult to construct problem cases in which rival ways of classifying possibilities have some claim to being called equally specific; besides, no alternatives are really basic, it is always logically possible to divide them further. Still, these criticisms of our first method only show that one gets into difficulties if one tries to use it generally; there would seem to be types of situation which are central in the study of games of chance, such as our card-drawing one, in which this requirement has a sufficiently precise meaning, and hence we might be able to use it in a definition of probability which would have at least some clear range of application. If, however, we ask what point there is in stressing basic alternatives or equal specificity where we can find them, it is obvious that the answer must be that basic alternatives are equally probable in the sense that we have no reason to expect one rather than another to be realized. Our first method, then, considered in the light of its rationale, boils down to the second. The classical definition, thus developed, will identify the probability of a possible event with the ratio of the number of 'favourable' possibilities to the total number of possibilities in a set of equiprobable possibilities. What is essential in this procedure is the use of the principle of indifference (or of insufficient reason) to determine equiprobability, that is, the claim that alternatives are equally probable if we have no reason to expect one rather than another. Now with some senses of 'probable', this principle would be simply false. E.g. with a limiting frequency sense: a die may be loaded, and we may not know this; if so, we have no reason to expect it to fall 6 uppermost rather than, say, 2 uppermost, and yet the limiting frequency of the former may be markedly higher than the limiting frequency of the latter. Again, with an objective chance sense: the fact that we have no reason to expect that some initial set-up will develop one way rather than another does not exclude the possibility that it should have a strong internal tendency in favour of one outcome rather than the other. But there *is* a sense of 'probable' in which the principle of indifference seems true (though not, I think, analytically true), and this is what I have proposed that the phrase 'simply probable' should be used to express, the good reasons end of the informal concept. If we have no reason to expect one alternative rather than another, we have reason to expect them equally —I mean to have equal expectations that either should occur, *not* to expect results like these to occur with equal frequency: that is quite another matter. Any other distribution of belief between the alternatives would be arbitrary. Admittedly, the application of this principle is unavoidably relative to a conceptual framework, just as the recognition of basic alternatives is. But still, given the powers of discrimination that we have, it will be reasonable for us to apply the principle in relation to them. Interpreted in this way, the classical definition might be defended not as the introduction of a new probability concept, but as a particular use of the concept of simple probability.[6] We have shown that in *certain* types of situation, where there is in fact a set of alternatives which, for all we know, are on an equal footing, we can graft a way of measuring probabilities and a consequential way of calculating with them on to the initially non-metrical concept of simple probability. If we have no reason to suppose that one card will be drawn rather than any other, it will be reasonable to have a stronger expectation that a red card will be drawn than that a king will be, and we can appropriately measure these justified degrees of belief by the corresponding ratios among the fifty-two equiprobable possibilities, 26:52 and 4:52 (or the fractions $\frac{26}{52}$ and $\frac{4}{52}$).

However, the classical concept of a ratio among possibilities comes to live a life of its own, the requirement that these should be in the simple sense equiprobable being taken for granted and no longer stressed or even mentioned, and it can then be called a distinct probability concept. But this is only the most elementary version of a range or measure concept: we shall come back to a more advanced version after looking at the logical relation concept.

[6] Laplace, op. cit., pp. 6–7 says 'The ratio of [the number of cases favourable to the event whose probability is sought] to that of all the cases possible is the *measure* of this probability' (my italics), but on p. 11 he slips into saying 'the definition . . . of probability . . . is the ratio of the number of favourable cases to that of all the cases possible'.

Before leaving the notion of equiprobability, however, I want to guard against a possible misinterpretation of the thesis, which in itself I accept, that it is only in certain types of situation, such as those found in what are called games of chance, that this method of introducing measures of simple probability can be applied. It is an empirical fact that certain sorts of things are suitable for use as gambling devices—cards, dice, and so on. It is an historical or sociological fact that they are so made as to be suitable for this: the cards in a pack are made all the same size and with the same design on their backs, within ordinarily detectable limits. Dice are tested by their manufacturers for lack of bias. We may well believe that, as a result, drawings of cards, throws of a die, etc., will have limiting frequencies, or objective chances, or both, which correspond to the probabilities calculated by the classical method. And it may be said that it is only because they do so that these are suitable types of situation for the application of the classical definition.[7] In other words, the classical method of grafting measurement on to simple probability involves a surreptitious appeal to frequency or objective chance. But this is not what I am saying. We do not first have to *find* that card-drawing produces, in the long run, an equal frequency for drawings of each of the fifty-two cards; we do not even first have to *suppose* that it will do so; and we do not have to determine thus, or in whatever other ways might be possible, that the objective chances of each card being drawn are equal. The mere fact that we know nothing that would lead us to expect one card rather than another to appear commits us to dividing our expectations equally between the fifty-two. What makes this type of situation suitable for the application of the principle of indifference is just that it contains a set of alternatives similarly related to our knowledge and ignorance, *not* that these alternatives occur with equal frequency in long runs of tests. We don't need equal frequencies or equal objective chances among the data for this procedure; and correspondingly, as I have said, we don't find them among the conclusions to which it validly leads.

The logical relation theory is commonly encumbered by a number of misleading suggestions and ways of speaking. Logical relation theorists have introduced such formulae as

[7] Cf. J. Venn, *The Logic of Chance*, ch. 4.

'*P*(*h*, *e*,)', to be read as 'the probability of hypothesis *h* on evidence *e*' or 'the probability of *h* given *e*'. Taken literally, these phrases suggest that once we have *e*, the probability in question attaches to *h*, to the hypothesis or conclusion on its own. But this is not what the logical relation account, in offering a distinctive probability concept, contends. As its name implies, it takes the central thing about probability to be some relation between *e* and *h*, not something that ever belongs to *h* on its own. The analogy used is that of entailment, considered as a relation of logical necessitation between the conjunction of the premisses and the conclusion in a valid argument. Between *e* and *h*, it is suggested, there is a relation of quasi-entailment, of partial support. Another problem lies concealed within the symbol '*e*'. What are we to take as an instance of such a formula as '*P*(*h*,*e*)'? One suggestion might be 'The probability that *x* has blue eyes, given that *x* is an Englishman'. But this will not do. '*x* has blue eyes' and '*x* is an Englishman' are not complete statements, but propositional functions. In any case there is no *logical* relation either between these propositional functions or between corresponding instantiations of them (e.g. 'Smith has blue eyes' and 'Smith is an Englishman') other than indifference. There may indeed be an *empirical* relation; e.g. there is no doubt some definite frequency, say 40%, with which the feature '. . . has blue eyes' occurs in the finite class either of now-living Englishmen or of all Englishmen dead, living, and still to be born. But if, on the strength of this, we say that the probability of '*x* has blue eyes' given '*x* is an Englishman' is 40%, the probability relation we are asserting is a contingent, empirical, one, not a logical relation. To get anything that has any plausible claim to being a logical relation we need to include the contingent frequency information as part of the evidence; the *e* that quasi-entails *h* must be at least something like '40% of Englishmen have blue eyes and *x* is an Englishman'—or, preferably, in order to get complete propositions in *e* and *h*, we should replace '*x*' in both '*h*' and '*e*' by some individual name such as 'Smith', and treat it *as* a name, not as a variable. Deductively speaking, the relation between '40% of Englishmen have blue eyes and Smith is an Englishman' and 'Smith has blue eyes' is still indifference: the truth, or falsity, of either would not entail the truth, or falsity, of the other. But there is now a formal relationship between the first statement and the second such that the first seems to give some support to the second, and that like support would be given by any statement of the same form as the first ('40% of *A*s are *B* and *C* is an *A*') to the corresponding statement of the same form as the second ('*C* is *B*'). There is room, here, at any rate, for a formally based relation of support.

The fact that e.g. 'Smith has a blue right eye' has a fairly high probability on the evidence that Smith is an Englishman, but a very low probability on the evidence that Smith is an-Englishman-with-a-brown-left-eye, cannot be used to support the logical relation theory of probability: the probability relations to which this contrast draws our attention are contingent, not logical ones.[8] It is a contingent fact that a brown left eye is unlikely to go with a blue right one: there might well have been a species in which, say, sexual selection based on odd aesthetic tastes had made it very common for individuals to have eyes of different colours. We cannot say that it is wrong to use such a formula as '$P(h,e)$' in this way, though it might be more perspicuous to use something which could be read as 'The probability of an Englishman having a blue right eye' and which might be written something like '$P(B(e))$', where this 'e' stands for the indefinite subject 'an Englishman' and the probability is ascribed not to a relation between propositions or propositional functions, but to the propositional function '$B(e)$'—i.e. 'an Englishman has a blue right eye'—itself. But whether we follow this suggestion or not, we must keep the contingent cases separate from the proposed logical relation ones. When we do so, we see that the contingent cases have no tendency either to support or to undermine the logical relation theory: they are just another matter, and the logical relation theory must be supported on other grounds, if at all.

In fact, such other grounds are easily found. Let us assume that some things are simply probable, in other words that we sometimes have good but not conclusive reasons for believing things. It seems clear that these reasons will often be, or be formulable as, other propositions that we know or believe. Also,

[8] This example is misused by H. Jeffreys, *Theory of Probability*, p. 15, in a way which lays him open to criticism by J. R. Lucas, *The Concept of Probability*, pp. 50–6, who in turn misuses this point as an objection to logical relation theories in general.

if these other propositions are not logically related to those for which they constitute good reasons—as, for example, 'Smith is an Englishman' is not logically related to 'Smith has blue eyes' —then there must be further propositions such that the conjunction of these with the reason-giving ones is logically related to those for which they give reasons—as the conjunction of '40% of Englishmen have blue eyes' with 'Smith is an Englishman' is related to 'Smith has blue eyes'. There must be some such formal relationship between the reason-giving conjunction and the conclusion, and this formal relation must somehow constitute a relation of non-deductive support: non-deductive, because the reasons though good are inconclusive, but support because these are, by hypothesis, reasons for believing something. It seems clear, too, that the non-deductive support must arise from the formal relation: if one body of evidence supports a certain conclusion, any other body of evidence which is formally related to another conclusion in the same way must support it too. In short, it seems undeniable that if there are good but not conclusive reasons for believing things, there must also be relations of quasi-entailment, of partial, non-deductive, support. And this kind of relation might well be called probabilification. There seems, then, to be a coherent way of introducing a relational probability concept, and one which we can hardly avoid once we have admitted the good reasons end of the informal concept. Is this concept, as is often thought, in danger of collapsing into a subjectivist one? Not in the most obvious sense of that accusation. As I have explained it, it is certainly not the concept *of* a merely subjective relationship, i.e. of one constituted by people's tendency to infer something from something else. Just as the concept of good reasons for believing makes an objective claim, so does the concept of probabilification which, I have argued, is carried along with it. Of course, a critic might argue that there is no such thing as objectively good reasons, that although talk about good reasons or likelihood or probability makes objective claims, these claims are never justified. That is, although this is a concept *of* something objective, a critic might say that it has in the end only a subjective status. And he might say the same about the supposed relation of probabilification; although in treating it as a formal logical relation and describing it as quasi-entailment, those who use this concept are implying that the relation holds objectively, the critic might argue that there *is* no such objective relation, that it should be explained away as a fiction based on people's tendency, in certain circumstances, to infer this from that. The logical relation theory is not itself a subjectivist theory, nor does it tend to turn into one; but a critic might argue that the supposed entities with which it deals have only a subjective status.[9] (It might be thought that the logical relation theorist would find it harder to defend his probability concept against this sort of external criticism than, say, a frequency theorist: but as we shall see at least one important version of the frequency theory is in much the same boat.) In this sense, and only in this sense, is there any force in the accusation that the logical relation theory tends towards subjectivism.

The most plausible candidate for the role of an example of quasi-entailment is the proportional syllogism, an argument whose premisses are of the form:

x % of *A*s are *B*
C is an *A*

and which has the corresponding conclusion of the form

C is *B*.

It seems plausible to say that these premisses support or probabilify this conclusion, that the nearer x is to 100 the more strongly they support it, so that x% can be used as a measure of the degree of probabilification: we can say that these premisses probabilify-to-degree-x% this conclusion.

It might be objected that such premisses probabilify such a conclusion only in the absence of other evidence, that *C*'s being *B* may be made more or less likely if we have other relevant information as well. There is a truth here, but it is confusingly stated. What is correct is this. As I shall argue later, when we consider 'detachment', we cannot derive the simple probability statement 'There is a probability of x% that *C* is *B*' from the stated premisses alone, but only from these subject to the proviso that they form the total available and relevant information, provided, that is, that we know nothing else that

[9] This criticism would be analogous to the view of moral qualities suggested in my 'A Refutation of Morals', *Australasian Journal of Philosophy*, 24 (1946), 77–90.

bears upon the question whether *C* is *B* or not. But this is a restriction upon detachment, upon extracting a simple probability conclusion from a probabilification, not upon the probabilification itself; the latter can be said to hold whenever we have the formal relation between premisses and conclusion (of the form '*C* is *B*', *not* of the form 'It is $x\%$ probable that *C* is *B*') set out above. In other words, '*P* probabilifies *Q*' means that '*P*' is formally related to '*Q*' in such a way that if we knew that *P*, but nothing else that had any bearing on whether *Q*, it *would* be simply probable (to the corresponding degree) that *Q*.

This proportional syllogism is, as I said, the most plausible formally complete example of probabilification: it is the one that lends most colour to the use of the term 'quasi-entailment', since where $x = 100$ we have, of course, a valid traditional syllogism. But the instances that conform straightforwardly to this pattern do not exhaust the cases where it is natural to speak of non-deductive support. Where we most urgently need a relation of partial support is between scientific hypotheses and theories and the bodies of evidence that are said to confirm them. It is significant that Carnap identified probability$_1$ with 'degree of confirmation'. There is here a topic of vigorous controversy. Some philosophers of science have tried to assimilate inductive support or the confirmation of hypotheses to the proportional syllogism type of argument; others have maintained that this cannot be done, and even that such confirmation is not a probability relation at all, in that it does not obey the standard probability calculus.[10] I cannot here enter into this controversy, but I mention it because it suggests that there might be at least two different, perhaps radically disparate, kinds of non-deductive support, one exemplified by the proportional syllogism, the other by the confirmation of hypotheses by testing. The logical relation concept of probability might need to be subdivided. Although we have, to start with, just one concept of partial support, it might turn out that we can and should develop in its place two more complex concepts, incorporating in each of them a more precise set of

[10] See e.g. D. C. Williams, *The Ground of Induction*; the works of Carnap and Popper and their followers *passim*; various papers in *The Problem of Inductive Logic*, ed. I. Lakatos; L. J. Cohen, *The Implications of Induction*, ch. 1; J. L. Mackie, 'The Relevance Criterion of Confirmation', *British Journal for the Philosophy of Science*, 20 (1969), 27–40.

rules, and that the rules for one of them would not satisfy the standard probability calculus.

On the other hand, it may be that what is needed is rather an extension of the kind of reasoning found in the proportional syllogism. For one thing, we may wish to use the pattern of argument where the *A*s of which $x\%$ are *B* are not actual things or events, but possibilities—hypothetical cases or occurrences, introduced by statements about what would happen if . . . or about the ways that things might be. But once we have made this leap, we can go on and look at a probability problem in terms of a realm of possibilities. Each possible situation can be characterized in as many ways as is relevant to the problem. Then within such a realm of possibilities, we can define the logical relation of probabilification, $P(h,e)$, as the ratio of the number of possibilities which satisfy both h and e to the number of those that satisfy e by itself. Or we may extend the notion further from a discrete to a continuous range of possibility, and measure $P(h,e)$ as that fraction of the area of possibility covered by e which is covered by h as well.

While in one respect this is an extension of the proportional syllogism, in another respect it is a development of the classical notion of a ratio of possibilities. And of course the problem of equiprobability, which that notion tends to sweep under the carpet, crops up again in a new form. Before we can speak with any precision of a fraction of an area of possibility, we must have some way of measuring such an area, we must be able first to assign a quantitative measure to the area of possibility in which e holds, and then to assign a measure on the same scale to that portion of this area in which h holds too.

It is this line of thought that Carnap pursued in order to give substance to the concept of a *logical* relation of probability.[11] His idea is that we first lay down a language in which the relevant possible situations can be described; this determines a set of 'state-descriptions', each of which fully describes one of the possible states of affairs which that language allows us to distinguish. We then assign measures to these state-descriptions in some systematic way; then this will determine in what fraction of the area in which some proposition e holds, some other

[11] *Logical Foundations of Probability*, chs. 3, 4, 5, esp. pp. 295–9; there is a simple exposition in A. Pap, *An Introduction to the Philosophy of Science*, ch. 12.

proposition h holds also. This looks promising as a way of generating determinate logical relations of quasi-entailment, but there are two basic difficulties.[12] One is that there seems to be some unavoidable arbitrariness about the assignment of measures, which can be illustrated by a very simple example. Suppose we consider a language containing just two individual names '*a*' and '*b*' and one predicate '*F*' and its negation 'not-*F*'. The 'universe' that this language can describe has just four possible states, with the state-descriptions '*Fa* & *Fb*', '*Fa* & —*Fb*', '—*Fa* & *Fb*', '—*Fa* & —*Fb*'. Let us use this to consider the inductive question: does the discovery (e) that a is F support the hypothesis (h) that b is F also? Now the most natural assignment of measures would be to give each of the four state-descriptions the same value, $\frac{1}{4}$. Then the area in which e holds covers two state-descriptions, so has the measure $\frac{1}{2}$. The area in which both h and e hold covers one state-description only, so has the measure $\frac{1}{4}$. Hence $P(h,e)$ would be $\frac{1}{4} \div \frac{1}{2}$, which is $\frac{1}{2}$. But this is the same as the initial probability of h, since it was true in two state-descriptions. So the observation that a is F would not increase the likelihood that b is F: inductive reasoning would be frustrated. But there is a simple way out. Think not only of state-descriptions but also of 'structure-descriptions'. This little universe has three possible structures: 'both objects are *F*', 'both objects are not-*F*', 'one object is *F* and one not-*F*'. Assign equal measures to each structure-description, and the four state-descriptions must have the values

$$Fa \;\&\; Fb : \tfrac{1}{3}$$
$$Fa \;\&\; -Fb : \tfrac{1}{6}$$
$$-Fa \;\&\; Fb : \tfrac{1}{6}$$
$$-Fa \;\&\; -Fb : \tfrac{1}{3}$$

It then follows that the measure of the area in which e holds is $\frac{1}{2}$, that of the area in which both h and e hold is $\frac{1}{3}$. Hence $P(h,e) = \frac{1}{3} \div \frac{1}{2} = \frac{2}{3}$, whereas the initial probability of h was still only $\frac{1}{2}$. So the observation that a is F raises the probability that b is F also: we can reason inductively after all.

But it then looks as if assignments of measures are arbitrary in an important way: we can make it that e raises, or does not

[12] Cf. Pap, op. cit., pp. 199–201 and 206–7; Carnap, op. cit., p. 565.

raise, the probability of *h* at will. *Logical* relations ought to be less elastic.

The second difficulty, a little harder to demonstrate but easy to anticipate, is of the same sort: the probability relations between propositions vary with the choice of a language in which to describe the universe, even with the mere addition of predicates that are not actually used in the propositions between which the relations are supposed to hold.

These well-known difficulties block the immediate establishment of logical relations of quasi-entailment by this method. But they do not show that the concept of probability as a measure of overlapping ranges of possibilities is itself incoherent. We may be able, in some cases, to justify a particular way of measuring these ranges; or we may be able, in some cases, to show that within some fairly broad limits the precise choice of an initial measuring system doesn't matter much; and once we have a measuring system that is *defensible in so far as it matters*, the range concept of probability can be used in a fairly determinate way.

Let us go back to the logical relation concept. I was dealing earlier with what we may call the strong concept of quasi-entailment or probabilification which makes this strictly a logical, *a priori*, relation. But I must concede that there is also a weaker relational concept. Suppose that the conjunction of two statements, *e* and *f*, would logically probabilify *h*, but that *e* alone would have no logical relation but indifference to *h*. (E.g., *e* might be 'Smith is an Australian', *f* might be '80% of Australians are city-dwellers', and *h* might be 'Smith is a city-dweller'.) Then we may say that *e* supports *h*; we are then, in effect, taking *f* for granted, and not explicitly mentioning the part it plays. We may even say here that the probability of *h* on *e* is 80%. Considered as a fact about *h* and *e* by themselves, this is clearly contingent: it reflects two truths, one contingent (that *f* holds) and one *a priori* (that *e* & *f* probabilifies *h* in the strong sense). Another possibility is this. Suppose that the conjunction of *e* and *g* would entail *h*, while *e* and not-*g* would entail not-*h*, but *g* is not known to be true, but has been given a certain degree of simple probability, say 60%, and is independent of *e*. (E.g., *e* might be 'Leeds will win their last two matches', *g* might be the conjunction of an appropriate true statement about the points system and the present points position in the League table with 'Arsenal will lose at least one of their last two matches', and *h* might be 'Leeds will be champions'.) Then again, we may say that the probability of *h* on *e* is 60% (in the illustration, that winning their last two matches would give Leeds a 60% chance of being champions). This too is only a contingent fact about *e* and *h*; again it reflects two truths, one contingent (that *g* has a simple probability of 60% and is independent of *e*) and one *a priori* (that *e* & *g* entails *h* while *e* & not-*g* entails not-*h*). In at least these two ways a relation between partial evidence and a conclusion or hypothesis may hold contingently, not *a priori*. We may call this the weak concept of probabilification.

The range concept of probability enables us neatly to bring together simple probability, strong, and weak probabilification. If we think of a universe of possibilities where both what is allowed as possible and some system of measures (e.g. but not necessarily, equal probability for each of a set of basic alternatives) are determined by the complete stock of available relevant information, the simple probability of a proposition will be that fraction of the total area of possibility within which it is true. The strong probabilification of *h* by *e* will be that fraction, of the area of possibility in which *e* is true, in which *h* is true also. The first sort of weak probabilification of *h* by *e* will be that fraction, of the area of possibility in which *e* & *f* is true, in which *h* is true also: obviously, if we cut down the original universe of possibility to that part of it in which *f* is true, this turns into the strong probabilification of *h* by *e*. The second sort of weak probabilification of *h* by *e* will be that fraction, of the area of possibility in which *e* is true, in which *h* is true also: that is, it will necessarily coincide, on this method of representation, with the strong probabilification of *h* by *e*, since *g*'s probability is independent of *e*. Also, if we cut down the original universe of possibility to that part of it in which *e* is true, the strong probabilification of *h* by *e* turns into the simple probability of *h* in this reduced universe; this is how detachment is represented in terms of the range concept.

Simple probability, relational probability, and the range concept are obviously very closely connected with one another. Nevertheless they should be regarded as distinct concepts. One particularly tricky point is that simple probability is *relative* but not *relational*. The simple probability that I ought to assign to a proposition depends upon what information I have and what I have not: it is relative to my state of mixed knowledge and ignorance. In saying 'Probably *P*' or 'It is $x\%$ probable that *P*' or 'There is an *n* to 1 chance that *P*' I am speaking from a certain point of view; but it is *P* itself that I am speaking about, not the relation between *P* and the knowledge-ignorance mixture that fixes this point of view. A simple probability statement is *close* to a statement of the form 'My total present state of knowledge and ignorance probabilifies *P* to such-and-such a degree'—but it is not quite synonymous with such a statement. The difference is, indeed, brought out by the range representation sketched above. The simple probability statement takes one's present state of knowledge and ignorance as setting the framework, as fixing the universe of possibilities under consideration. A statement that explicitly refers to this present state treats it as only a subdivision of a wider area of possibilities. But we can move easily, and, if the range representation is defensible, validly, between the two. The range concept elaborates both the simple concept and the logical relation one, and tries to provide more of a metric than either could generate on its own; but it is more enterprising and therefore more exposed to criticism than either of the others.

§ 3. *Frequency*[13]

There are clearly two frequency concepts of probability. The first, that of frequency in a finite class, is unproblematic. If the class of *A*s is finite, it has a definite number of members, and the proportion of these that have the characteristic *B* will also be definite (though, of course, it may or may not be known). Also, probabilities interpreted as such proportions will obey the standard calculus. The second concept, of limiting frequency in an infinite or indefinite sequence, is more difficult. It is meant to apply primarily to empirical sequences, such as tosses of this penny. One problem is whether the notion of convergence to a limit, which is at home in series determined by some

[13] See particularly R. von Mises, *Probability, Statistics, and Truth*, and H. Reichenbach, *The Theory of Probability*.

mathematical formula, can be applied to an empirical sequence. Another problem is whether the requirement of randomness can be coherently applied to an empirical sequence. I shall comment briefly on these problems.

First, if there were an empirical series which actually did continue indefinitely, then it could in fact exhibit a limiting frequency. Suppose, for simplicity, that a penny never wore away, and was tossed in the usual way, or the usual variety of ways, at intervals of ten seconds *for ever*. The results of such tosses would form an indefinitely extended sequence of *H*s and *T*s. Let us suppose that the proportion of *H*s in the first ten tosses is $\frac{4}{10}$, in the first hundred $\frac{56}{100}$, in the first thousand $\frac{542}{1000}$, in the first ten thousand $\frac{5453}{10000}$. It is surely possible, though in no way necessary, that the following should hold: 'For any number δ, however small, there is some number of tosses, N, such that for every n such that $n > N$ the proportion of *H*s in the first n tosses lies somewhere between $0{\cdot}545+\delta$ and $0{\cdot}545-\delta$.' And if this held, it would be correct to say that 0·545 was the limiting frequency of *H*s in this sequence. That is, a concrete empirical sequence, indefinitely extended, could in fact have a limiting frequency. Of course we with our finite lives and limited powers of observation could not observe or conclusively verify this, or conclusively falsify it either, but that is neither here nor there: the meaningfulness of a statement does not depend on the possibility of checking it.

Secondly, there is randomness, which von Mises made a defining characteristic of his 'collectives', then identifying probability with a limiting frequency in a collective. He defined randomness as insensitivity to place selections: a sequence of, say, *H*s and *T*s is random if and only if every sub-sequence formed from the original sequence in any systematic way (e.g. taking every second member, taking every member that follows two successive *H*s, and so on) has the same limiting frequency of *H*s as the main sequence. It has often been alleged that this requirement is too strong and leads to contradictions, and frequency theorists have themselves been prepared to settle for a somewhat weaker definition or even to do without this requirement altogether.[14]

[14] See particularly K. R. Popper, *The Logic of Scientific Discovery*, pp. 154–80; also Hilda Geiringer, 'On the Foundations of Probability Theory', *Boston Studies in the*

We must concede, of course, that a sequence cannot be insensitive to all place selections whatever: any sequence containing *H*s and *T*s with a limiting frequency for *H*s between 1 and 0 must have sub-sequences that contain only *H*s and others that contain only *T*s, indeed it must have sub-sequences with every frequency for *H*s between 1 and 0. We cannot demand insensitivity to place selections made with knowledge of the results. But can we demand insensitivity even to all place selections that could be specified in advance? If the result of each toss is causally determined, in however obscure a way, a place selection could be made which prescribed, before the penny had come down in each toss, whether that toss was to be included or not; if such a selection took account of the immediate causes of the result at each toss (e.g. how strongly the coin was spun) it could yield a sub-sequence that had, say, a much higher frequency for *H*s than the main sequence. If we want a sequence to be insensitive to this sort of place selection, we must choose one in which the result of each trial is produced indeterministically at the final stage. Alternatively, we could content ourselves with insensitivity to all place selections determined by mathematical formulae which, in deciding whether each toss is to be included in the selection, take account only of its position in the series of tosses and of the results of earlier tosses. An infinite series, unlike a finite one, might be random in this sense: though, as we noted, there would *be* discrepant sub-sequences, they might not be describable by any formulae.

Discussion of this kind of randomness requirement has been confused by several errors. Although a formula can in principle be found that will describe any *finite* sequence of, say, *H*s and *T*s, a strictly infinite sequence could obey no formula at all. Again, Popper finds a contradiction between the requirement that the frequency should converge to a limit, which he thinks must involve obedience to a law, and the complete absence of lawfulness demanded by von Mises's definition of randomness. But convergence as described above (though I have mimicked the mathematical definition) can be a matter of brute fact, governed by no law. Verificationist misconceptions tend to intrude here.

Philosophy of Science, Vol. III, ed. Cohen and Wartofsky, pp. 212–14; Reichenbach, op. cit., ch. 4.

We should need a law to *prove* that the series would converge, but a concrete sequence could *in fact* converge to a limiting frequency without our being able to prove this, and without violating the extreme randomness requirement. Also, Popper quotes from Kamke the view that we need to prove that there are collectives, and the obvious way to do this is to construct some; but of course we cannot construct an *infinite* sequence except by prescribing some law which it is to obey. But for one thing there might well be collectives even if we couldn't prove that there are. For another, we can prove that there are some without constructing any. As Lucas says, following Church,[15] 'there are $2^{\aleph_0}$ possible infinite series that might constitute a von Mises collective, but only $\aleph_0$ for which the rule specifying them could be antecedently given in a finite number of terms'. And we need not be disturbed by Lucas's remark that, thus defended, 'Frequency theory has become a mathematician's pastime instead of a plain man's guide'. The plain man can *use* the frequency concept even if he needs the help of mathematicians to *ensure* that it has application. There is some vagueness in von Mises's specification of the complete set of place selections to which a random sequence is to be insensitive, but his followers have formulated more exactly what he seems to have intended, and have shown that sequences can be both convergent and random in this sense.

Still, it is relevant to ask for what purposes randomness is required. One purpose is that a sequence should be sufficiently free from regularities to enable us to prove that Bernoulli's Central Limit Theorem (see § 8 below) holds for it. For this purpose, something much less stringent than von Mises's requirement will do, and Popper has shown how to give a constructive proof that there are sequences which are random enough to let this theorem be proved for them. Another reason for wanting randomness would be this: as we shall see in § 7, we can infer a simple probability for an individual result from a limiting frequency provided that the state of knowledge and ignorance for which this simple probability holds does not include any information that bears distinctively upon this individual result; complete randomness will in general ensure the

[15] Lucas, op. cit., p. 99, referring to A. Church, 'On the Concept of a Random Sequence', *Bulletin of the American Mathematical Society*, 46 (1940), 130–5.

necessary degree of ignorance, whereas anything less might fail to do so. I think that von Mises's own purpose was different again but related to this. It is only for a sequence which is completely random that a probability description has the last word. If there were any even hidden regularity, then a description in terms of the limiting frequency would be only an interim, makeshift account of what was going on. Pressed to its extreme, this line of thought implies that only indeterministic sequences —i.e. ones such that an indeterministic factor enters at some stage (not necessarily the last) in the process that generates each result in turn—will be strictly random and will be ultimately and adequately described in terms of their limiting frequencies. But there may be such sequences: there is no internal inconsistency in the kind of description that von Mises offers.

On the other hand, I must admit that this way of defending the limiting frequency concept takes us well away from actual experience: limiting frequencies can never be observed. The statement that tosses of a certain penny have a certain limiting frequency of heads is at best an hypothesis; it has at best the same status as a universal law statement or law-like statement about an individual object, e.g. that whenever this penny is placed in water it sinks. But the limiting frequency statement is often also hypothetical in another way—whether a law-like statement is analogous in this respect too is a matter of dispute. We are not really concerned with any actual sequence of tosses of this penny. Real pennies, unlike the one we supposed above, are never tossed more than, say, a million times. Moreover, they do wear, and it is reasonable to suppose that wear may change our unbiased penny into a biased one or vice versa, i.e. alter its limiting frequency of heads. For these reasons, the limiting frequencies we are concerned with are not, as suggested above, hypothesized frequencies in actual sequences, but hypothesized frequencies in hypothetical sequences: the sequence we are concerned with is what we *would* get *if* this penny were tossed indefinitely *without its present characteristics being changed.*

This does not make the limiting frequency concept illegitimate or obscure; nor does it compel us to transfer it from the realm of ordinary empirical discourse to that of pure mathematics. On the other hand, it does weaken the claim that this is a concept of an *objective* probability. For although, as I have argued, there could be an objective limiting frequency in an objectively random collective, this is not what we are usually talking about. What we are talking about in most actual applications of the limiting frequency concept is *what would happen if* . . . In view of what I say about conditional statements,[16] I would say that the concept of limiting frequency is not in general the concept of something fully objective, although von Mises and most other frequency theorists intended it to be so. It may turn out, then, that the limiting frequency theory is after all exposed to the charge of tending towards subjectivism in just the same way as the logical relation theory. (By contrast, the frequency in a finite class is impeccably objective.) This does not mean, however, that statements about limiting frequencies are arbitrary. There can be evidence that tells for or against them, they can be confirmed or disconfirmed; in a concrete situation, one judgement about a limiting frequency may be reasonable and others highly unreasonable. In an elementary way, it is obvious in what sense a frequency in an actual sequence may look as if it were going to converge to a limit in the neighbourhood of a certain value: we know what it means to say 'It looks as if the frequency of heads in the sequence of tosses of this penny is settling down somewhere around 0·5287'.[17] In a more elaborate way, it may be possible to use Bernoulli's Central Limit Theorem or one of Bayes's Theorems to show how the observation of frequencies in finite runs confirms an hypothesis about a limiting frequency. I shall consider this further when I come to discuss transitions between different concepts of probability. But it is important to note that *if* this procedure works, it will be (among other things) the confirming of *an hypothesis about a limiting frequency* by *observations of frequencies in finite groups*. The limiting frequency is on the hypothetical, not the observational, end of the confirmation. It is wrong to suggest that Bernoulli's Theorem would establish a link between probability and *limiting* frequency which the frequency theory lays down by definition instead; even after probability had been defined as limiting frequency we should still need Bernoulli's Theorem or Bayes's to provide the link between this and observed finite class frequencies. Defining probability as limiting frequency leaves highly fallible the judgement about

[16] Ch. 3 above. [17] Lucas, op. cit., p. 110.

what value either of these has in any concrete case. There is, indeed, a procedure (that of R. B. Braithwaite) in which probability is defined by a rejection rule that uses finite sets of observations: roughly, the *meaning* of the hypothesis that the probability of an *A*'s being *B* is $x\%$ is that this hypothesis will be rejected if the observed proportion of *A*s that are *B* in a finite observed set of *A*s differs too much from $x\%$.[18] But this procedure is very different from that which defines probability as limiting frequency in the sense outlined above. Braithwaite's procedure *would* give a definitional link between probability and observed frequencies in finite groups, and so make the appeal to any theorems unnecessary. But the identification of probability with limiting frequency in the ordinary sense leaves the link with observed frequencies still to be made, and I shall discuss this further in § 7.

§ 4. *Objective Chance and Propensity*

'Objective chance' is perhaps the most elusive of all our five concepts. It is hard to say precisely what an objective chance would *be*, but we can creep up on it by explaining what it is supposed to *do*. Objective chance is a counterpart of the power or necessity in causes for which Hume looked in vain.[19] Just as the power in a cause would be something present in every instance of a certain kind of cause which somehow *guaranteed* the subsequent occurrence of the corresponding effect, so a penny's having, at each toss, a certain chance of falling heads and a certain (perhaps different) chance of falling tails would be something present in the initial stages of every individual tossing process which *tended* to produce the result 'heads' and *tended* to produce the result 'tails', where these tendencies might be either equal or unequal. A clearer example might be a four-sided top which, when it stopped spinning, would lie with one of the four sides (marked '1', '2', '3', and '4') uppermost. Each spin of this top might have a certain *chance-distribution*, say 0·3 for side 1, 0·26 for side 2, 0·2 for side 3, and 0·24 for side 4. That is, the tendencies in this case would be nearly but not quite equal, the tendency for the top to come to rest with side 1 up

[18] R. B. Braithwaite, *Scientific Explanation*, ch. 6.

[19] D. Hume, *A Treatise of Human Nature*, Bk. I, Part III, Section 14.

would be the strongest, with side 3 up the weakest, and the others in between. (The chances in such a chance-distribution would of course add up to 1.) Such a chance-distribution would *tend* to produce the various results, but of course would not *guarantee* any of them. It would, presumably, guarantee a distribution of *limiting* frequencies in an indefinitely extended sequence of spins corresponding exactly to the distribution of chances, provided that the top and the way of spinning it did not change. And, in view of Bernoulli's Theorem, the proposed chance distribution would *tend very strongly* to produce distributions of frequencies in long finite runs of spins that were fairly close to the distribution of chances.

Of course this account is still obscure, because the word 'tend', on which it relies, cries out for further analysis. We are caught in something like the circle that, as Hume pointed out, we fall into when we try to define any one of the terms 'efficacy', 'agency', 'power', 'force', 'energy', 'necessity', 'connexion', and 'productive quality' by means of any of the others.[20] However, it may make things a bit clearer if we say that objective chance, in the primary and strongest sense, would be an indeterministic counterpart of causal necessity. The notion as applied to our top would be as follows—few thinkers, in fact, would want to apply this notion to a top, as we shall see, but I use this as a simple illustration. Consider a top, like the one mentioned above, whose behaviour is strictly indeterministic. It is given, on each trial, an exactly similar initial spin, the surface on which it spins is uniform and unchanging, and so are any causally relevant surrounding conditions. That is, the initial conditions of each spinning trial are exactly alike. Yet, being indeterministic, this top does not land in the same position at the end of each trial. It lands sometimes one side up, sometimes another, with frequencies which, in finite runs, look as if they were converging to something near the values in our suggested chance-distribution, 0·3 for side 1, 0·26 for side 2, 0·2 for side 3, and 0·24 for side 4. The chance-distribution, then, is a postulated something or other in each trial which produces these different results.

This illustration is rather fanciful, because we do not generally believe that tops are indeterministic. But many physicists say

[20] Loc. cit.

that atoms and sub-atomic particles are indeterministic: if so, this strong notion of objective chance *might* have application to them. For instance, it might be that each atom of some radioactive isotope was in the *same* initial condition, but that when subjected to the simple trial of being left alone for a certain length of time some atoms decayed and others did not, the proportion of the atoms in any collection which decay in time t looking as if it was converging to a limit, say $1-(1/2^{t/T})$ (where T is what is called the half-life), as the number of atoms in the collections was increased; in these circumstances we might be tempted to ascribe to each of these similar atoms the same *chance* of decaying in time t, measured by this value $1-(1/2^{t/T})$ and thinking of this *chance* as a something or other present in each atom which somehow produces the observed pattern of proportional decay.

Even in this application, however, the notion of objective chance that I have sketched may seem fantastic. Is there anything less extreme which will nevertheless give us a distinctive concept of this sort? Let us go back to the spinning top and consider a more plausible account of what actually happens to it. The top is physically nearly but not quite symmetrical. It does not receive an equal angular momentum at each spinning trial, but is sometimes given a stronger push, sometimes a weaker one: what we can call the *spinning inputs* vary in an at least apparently random way over a considerable range. Because of the slight asymmetry, more of the inputs within this range will (deterministically) lead to the result '1 up' than to the result '3 up', and so on. There is in principle a causal explanation of the actual frequencies of results in any long finite series of spins (as well as of each individual result). The explanation of the frequencies is given by the physical asymmetry and the actual distribution of spinning inputs together. A very crude illustration would be as follows. In 100 trials, the spinning input has 100 different values, which we shall merely number for reference from 1 to 100. Inputs with values 1 to 6 result in side 1 coming up, from 7 to 11 in side 2 coming up, from 12 to 15 in side 3 coming up, from 16 to 20 in side 4 coming up, from 21 to 26 again in side 1 coming up, and so on. The asymmetry of the top means that more possible inputs will result in 1 uppermost than in 3 uppermost, and the actual distribution of inputs then produces a frequency for each result, in the series of 100 spins, which is close to the value for that result in the chance-distribution.

Can we say that there is an objective chance-distribution in this sort of case? We can, if and only if the range of spinning inputs itself results from a chance-distribution in the indeterministic sense explained above. An objective chance input, interacting *deterministically* with the physical characteristics of the top, will give each spinning trial an objective chance-distribution. The laws of working of the top itself may be deterministic, but each spin will have only an objective chance of producing each possible result if there is only an objective chance that this or that input will be supplied. To get an objective chance, we do not need indeterminism in the spinning process itself, but we do need indeterminism *somewhere*. If, alternatively, the spinning inputs themselves are fully determined in some strictly causal way, then we cannot say that each trial has an objective chance of producing this or that result. What result each trial produces is open to a complete causal explanation and is not objectively a matter of chance at all. Of course, we may still *speak* of chances and a chance-distribution here, but such talk will reflect our partial ignorance of the spinning input in each individual trial and of the way it will interact with the physical asymmetry of the top. If we ascribe chances to *individual* trials in these circumstances we ought to mean only something that could be analysed in terms of either the simple (or informal) or logical relation concept, and if we ascribe chances to a finite *set* of trials we ought to mean that these are, or that we expect, or are justified in expecting, the corresponding frequencies for the different results. In other words, a distinctive concept of objective chance, which would not collapse into one of the other concepts, can be used only where the user is implicitly or explicitly claiming that there is some indeterminism at work, either in the trial process itself or in some causal ancestor of some input into that process.

Although we cannot ascribe objective chances without at least implicitly accepting indeterminism, accepting indeterminism does not immediately force us to assert objective chance in the strong sense. Prima facie at least, there might be an indeterministic top as described above, one which just did exhibit, in finite runs of trials, frequencies for the various results which looked as if they were converging towards limits, and yet there might not be any constant intrinsic chance-feature of each trial's initial conditions which *produced* this pattern of results. Of course there will be other constant features of these initial conditions—the physical characteristics of the top, the angular momentum supplied, and so on—but none of these need be a chance or chance-distribution. Whether we can coherently speak about a statistical *law* without ascribing objective chances is a further question, which I cannot discuss here. But on the face of it we are no more committed, by the recognition of a scatter of results with frequencies that look as if they were converging to limits, to ascribing chance-features to each trial, than we are committed, by the recognition of a uniform set of results that look as if they instantiated a strict causal law—the penny always sinks in water—to ascribing an intrinsic power or necessity to each individual instance of the cluster of causal factors.

This suggests another weakened concept of chance. We might—indeed we often do—in words ascribe a chance or chance-distribution to each individual trial without seriously meaning to assert that each trial has in it some indeterministic counterpart of a power, but merely as an indirect way of saying that an indefinitely repeated sequence of exactly similar trials would exhibit the corresponding limiting frequency. Here the ascription of chances to individual cases is only a manner of speaking, all we are seriously asserting is a limiting frequency in a hypothetical sequence. This, therefore, is not, strictly speaking, another form of the objective chance concept, but the limiting frequency concept again in a verbal disguise.

There is, then, only one concept of objective chance which remains distinctive and does not collapse into one of the other concepts we have already surveyed, and that is the strong, perhaps even fantastic, concept with which I began, the indeterministic counterpart of the concept of power which Hume said to be part of our concept of cause, but for which he could find no source in any observation of objective causal sequences, and so had to explain as a projection of a 'determination of the thought'.[21]

[21] Loc. cit.

I have admitted that this is an obscure concept. Chances, like powers, are not only unobserved; we cannot even begin to say what it would be like to observe them. Yet I do not think that we can rule these concepts out of consideration on that account: it may be legitimate and meaningful to postulate a something-we-know-not-what as that which produces or would explain something that we are acquainted with.

The propensity view of probability is a development of the notion I have been calling objective chance. Sir Karl Popper, indeed, seems to identify a probability propensity with what I call an objective chance, and a propensity-distribution with a chance-distribution, while insisting, correctly, that either of these would be a dispositional property of the total experimental set-up—e.g. the die, the way it is thrown, and the gravitational field.[22] A more elaborate view is that of D. H. Mellor, who uses the term 'chance' much as I have been using it, but uses the term 'propensity' to refer to a dispositional property not of a particular trial or set-up but of something that enters into a trial or set-up: the bias of a loaded die is a propensity, a standing dispositional property of the die of which a certain chance-distribution when it is thrown in a certain way is a manifestation or display. So is the half-life of a radioactive atom: the display of this propensity will be its chance of decaying when left alone for a certain time.[23]

What I have said about dispositions and dispositional properties has a bearing on both these views.[24] Suppose first that the propensity and the chance are identified, that is, that the objective chance itself is treated as a disposition. Then the display of that disposition will be the corresponding frequency in the results of trials. Then if this disposition were merely what I have called a minimal disposition, it would consist simply in the holding of the appropriate conditionals of the form 'If the trials are repeated indefinitely there is such-and-such a limiting

[22] K. R. Popper, 'The Propensity Interpretation of Probability', *British Journal for the Philosophy of Science*, 10 (1959), 25–42, and 'The Propensity Interpretation of the Calculus of Probability', *Observation and Interpretation in the Philosophy of Physics*, ed. S. Körner, pp. 65–70; cf. W. Kneale's remarks in discussion reported, ibid., pp. 79–80.

[23] D. H. Mellor, *The Matter of Chance*, ch. 4. Another version of the propensity theory is given by R. N. Giere, 'Objective Single Case Probabilities and the Foundations of Statistics', to appear in *Proceedings of the Fourth International Congress on Logic, Methodology, and Philosophy of Science.*

[24] Cf. ch. 4 above.

frequency in the results'; that is, this view would collapse into the account already given of a hypothetical limiting frequency. (This frequency display is, as we have seen, the result of the repeated operation of a like degree of chance in each of a large number of trials. The various possible results of an individual trial, each of which the set-up merely tends more or less to bring about, are all that we could call the display-in-a-single-trial. Consequently no minimal dispositional account is available for the chance in a single trial: for this, we should need conditionals of variable strength, of such a form as 'If *P* *x*-per-cent-then *Q*', and these would have to be invented for this purpose.) If we move to what I have called the realist view of dispositions, we add merely that the set-up which has the propensity is *such that* there is this hypothetical limiting frequency, that is, that there is some categorical, non-dispositional feature of the set-up which contingently, by an indeterministic quasi-causation, gives rise to this frequency pattern in the results—e.g. the physical asymmetry of our indeterministic top quasi-causally produces the unequal frequencies for side 1, side 2, and so on. But still the propensity terminology introduces no further entities apart from the ordinary physical shape, mass-distribution, etc., and the hypothetical limiting frequency. Only if we go on to what I have called the rationalist view of dispositions will the chance-propensity constitute an addition to the picture of probability given by the hypothetical limiting frequency account. This view will postulate a further feature of the set-up in each trial, over and above all the ordinary features. This feature quasi-entails, with the appropriate degree of probabilification, each possible result at each individual trial. Only on this view of dispositions does the propensity remain a feature of the individual trial as opposed to the series of trials. But also, thanks to Bernoulli's Theorem, these quasi-entailments at each trial are so integrated that in the limit the propensity actually *entails* the corresponding limiting frequencies. The ordinary physical features clearly enter into neither this quasi-entailment nor this entailment, so this new feature must be distinct from them all. This is indeed, therefore, a distinctive theory, one which postulates an additional member of the array of probability facts and phenomena. But it is open to the decisive objections brought against this rationalist view in Chapter 4. The alleged new feature is logically related (by entailment or quasi-entailment) to each individual result and to what would be the total pattern of results in an indefinitely extended series of trials. But it is a distinct occurrence from all of these: it is present in the initial stages of each trial, before the result has come about. But there can be no such logical connections between distinct occurrences. Nor is there any good reason for postulating this extra feature to help to explain how the physical features of the set-up bring about the pattern of results. Admittedly this is something that is hard to understand because it is, by hypothesis, an indeterministic bringing about. But if we insert the chance-propensity which explains all too well, by logical necessity, the pattern of results, we have still as hard a task as ever to explain how the physical features give rise to this propensity: the problem has merely been shifted and obscured, not solved.

Suppose, alternatively, that the propensity is distinguished from the chance, and the chance itself, not the frequency, is taken as the 'display' or 'manifestation'—in a slightly odd sense, since a chance is very far from manifest—of the propensity. The main problem is now about the chance. If this is itself a disposition, all the points made above still apply; but if it is not a disposition it is just an obscure something-or-other. In any case, the supposed propensity of which this chance is the display, being a disposition, in turn faces the choice between minimal, realist, and rationalist interpretations; only on the last of these will it add anything to the ontology of probability, and again the decisive objections apply: it is supposed to be a distinct occurrence, now, even from the chance, and yet to be logically related to it; and it is an extra entity, gratuitously postulated, which does not really help to explain what goes on.

But of course these criticisms are directed against the propensity account as an ontological theory. I have no objection to a propensity style of description, as long as it is understood that in speaking, say, of the propensity of our indeterministic top to favour side 1 we are asserting no more than that there would be such-and-such a set of unequal limiting frequencies for the various sides coming up and that this arises somehow (indeterministically, by a contingent quasi-causation) from the physical asymmetry of the top.

I think, therefore, that there are strong reasons for doubting whether either form of propensity theory has ontologically valid applications. Also, by bringing out into the open, and attacking, one sort of thing that an objective chance might be, these criticisms cast doubt on the applicability of the concept of objective chance itself. But they do not go so far as to show that concept to be incoherent: it can be so developed as to be distinct from all our other probability concepts, and this distinct concept *might* have real applications, though I doubt whether it has.

§ 5. *Propositional Functions*

While discussing the logical relation concept, I mentioned that probabilities can be ascribed to propositional functions: we may say 'The probability of *an* Englishman having blue eyes is 40%' or 'The probability that *a* radium atom will disintegrate if left alone for 1590 years is 50%' and so on, and we may write these in some such form as 'The probability of $F(g)$ is x%'. To get the first statement we take 'g' as a variable ranging over Englishmen, 'F' as the predicate 'has blue eyes' and 'x' as 40; to get the second we take 'g' as a variable ranging over radium-atoms-left-alone-for-1590-years, and 'F' as 'disintegrates', and 'x' as 50. This propositional function formulation is a very natural way of treating at least some probability statements.[25] Does it point to a further probability concept, additional to those we have surveyed already? I think not. I think that this way of speaking represents one or other of the previous concepts. Thus the first statement might mean 'If something is known simply as an Englishman, it is reasonable to give the degree of belief measured as 40% to the proposition that this something has blue eyes'. This is a conditional statement using the simple probability concept with some metrical extension. Again it might mean 'There is some true statement the conjunction of which with any statement of the form "x is an Englishman" probabilifies to degree 40% the corresponding statement of the form "x has blue eyes" '. This uses the strong logical relation concept. What amounts to the same thing can be expressed in terms of the weak concept of probabilification: 'Any statement

[25] Cf. Lucas, op. cit., ch. 4.

of the form "*x* is an Englishman" probabilifies to degree 40% the corresponding statement of the form "*x* has blue eyes" '. Again, our original statement might mean '40% of Englishmen have blue eyes', which is a frequency statement about a proportion in a finite class. It is barely conceivable that it might mean 'The genetic lottery that produces Englishmen has a 40% objective chance of producing one with blue eyes'. Of course, it might mean some mixture of these. Since the first four may well all be true together, the propositional function formulation may be conveniently ambiguous between the four. Similarly, the statement about the probability that *a* radium atom will disintegrate might mean that if something is known only as a radium atom left alone for 1590 years it is reasonable to have equal expectations that it will decay and that it will not; or again that there is the appropriate logical relation; or that there is a limiting frequency of 50% in the class of radium atoms left alone for 1590 years arranged somehow as a series (it must be a *limiting* frequency, since radium atoms seem not, like Englishmen, to form a closed class); or that each radium atom thus left alone has a 50% objective chance of decaying. Again, the meaning of such a statement might well be a mixture of some or all of these: all four might be true together, or might be thought to be so, and the propositional function probability formula might therefore be conveniently ambiguous between them. But I don't think that the propositional function formula points to any additional probability concept: its use is to be understood in terms of one or more of the other concepts. The force of the indefinite article in '*an* Englishman', '*a* radium atom', etc., is to ascribe some feature (e.g. an objective chance) to every member of the class or to relate something (e.g. some degree of reasonable belief) to anything known only under the description 'Englishman', 'radium atom', or whatever it may be; only by an extension can it be used to ascribe something (e.g. a frequency) to the class itself.

§ 6. *Meaning Shifts*

We seem, then, to have five main probability concepts: the informal one, which hovers between the extremes of guarded assertion and good but not conclusive reasons for believing—

I have called the latter extreme simple probability; the range concept, of which the classical concept is an anticipation; the logical relation one, which may need to be subdivided into that exemplified by the proportional syllogism and one of confirmation; the frequency concept, which certainly needs to be subdivided into frequency in a finite class and limiting frequency; and the concept of objective chance. I hope that the discussion so far has shown that despite some indeterminacy and some obscurity, none of these concepts can be ruled out on grounds of poor internal health. It seems to show that (though there are very close links between the first three) none of the five can be reduced to any of the others. And I think it is obvious that every one of these concepts is actually in use, every one of these accounts crystallizes some of our ways of thinking and speaking about probability. This point may be brought out by noting the different kinds of subjects to which these different probability concepts are applied. The informal concept assigns probability to a proposition or statement, or to a that-clause; what is said to be probable is a certain complete judgement or belief or conclusion; but this is said to be probable in a context of incomplete knowledge. The logical relation account finds not so much probability as probabilification, and ascribes this to an argument or argument form, to an ordered pair consisting either of a set of premisses and a conclusion or of a body of evidence and an hypothesis. The range concept allows for ascriptions of probability to subjects of both these kinds. The simple frequency account ascribes probability to a different sort of ordered pair, a predicate or feature and an actual finite class; the limiting frequency account ascribes it to a predicate or feature and (in general) a hypothetical class, the merely possible indefinite extension of some empirical sequence. Objective chance is ascribed to the early stages of each individual process of a certain kind.

But though these concepts are not reducible to one another they are related; it is natural for words of the probability group to creep on from one of these senses to another. In fact, what I have called the informal concept is the central one from which the others are easily developed by shifts and extensions.

Using the informal concept, we are saying or hinting that some belief has some objective justification, that anyone in the appropriate circumstances ought to give the proposition in question some degree of credit. If so, there must be some reasons which partially support this belief, some information in someone's possession which makes the belief probable. And it is at least plausible to suppose that this can be so only if this evidence is formally related to the belief in some special way that constitutes this support. The informal concept thus leads on to the concept of probabilification. If evidence makes a conclusion probable-in-the-informal-sense, then this evidence must make-probable-in-the-logical-relation-sense that conclusion; and it is natural to use terminology derived from the word 'probable' to describe the relation to which our attention is thus drawn. The probability talk shifts from the conclusion which is partially supported to the partially supporting relationship.

I have argued that the classical concept of a ratio among possibilities could be viewed merely as a quantitative extension of the informal concept, a way of building measured probabilities of the simple sort out of the judgement that (in the light of the principle of indifference) certain possibilities are such that we have equally good reasons for expecting each of them. But we could take the classical definition just as a definition, as identifying probability simply with a ratio among possibilities. This ratio, in any concrete instance, will be *part* of what supports the simple probability judgement—only part, for there is the other vital element that these possibilities are, in the sense explained, equiprobable. To apply the probability terminology to the ratio of possibilities, then, represents a bigger shift than in the logical relation use: the shift is now from the conclusion to part of the evidence, not merely from the conclusion to the relation between evidence and conclusion.

The range concept requires for its exposition a half-reified realm of possibilities. If we see through this manner of representation, we come back to the simple or to the relational concept or to both together; but like the classical concept of which this is a development, this manner of representation tends to take on a life of its own. We have then a *projection* from simple probability and/or relational probability treated as a description of an imaginary realm.

A shift of meaning like that which identifies probability with a ratio among possibilities could be said to occur also when probabilities are identified with frequencies. If 80% of Australians are city-dwellers, and all we know about Smith is that he is an Australian, it does seem (for reasons about which I shall say more later) that this limited information makes it fairly reasonable for us to believe that Smith is a city-dweller. The latter conclusion, then, is simply probable or has a fairly high informal probability; we then transfer the term 'probability' from this conclusion to the most prominent item in the information that makes this conclusion probable, and call the statistical fact that 80% of Australians are city-dwellers a probability. But this shift can also be looked at in another and historically more correct way. In the study of games of chance we establish various probability values by the classical method —that the probability of throwing a six with a die is $\frac{1}{6}$, of throwing a total of exactly nine with two dice thrown simultaneously is $\frac{1}{9}$, and so on. Now if we have philosophical doubts about the classical interpretation and justification of these values, but note that they are somehow pragmatically acceptable, we shall look around and try to find some real and empirical quantities that have these values. And of course we find that frequencies in long finite runs do have roughly these values, and with some ingenuity we can construct hypothetical limiting frequencies that have exactly these values. So we are tempted to shift the term 'probability of x%' so that it applies to a class of entities that have just such percentages built into them. The philosophical motivation for this shift is that frequencies have (roughly) the right values and obey the right calculus to be described in the ways in which probabilities are already described, and that we doubt whether any other philosophically respectable entities have these. Putting it very crudely, what we call probable usually happens in the long run, we can't find anything else that being probable might be within the ontology to which we confine ourselves, and so we identify being probable with usually happening in the long run.

The notion of objective chance may be reached by a shift of a different kind. Just as, according to Hume, we construct the notion of necessity by projecting our expectation on to the objective causal sequence, so (a sceptic would say) we construct the notion of objective chance by reinterpreting the reasonableness of a belief (in the light of partial knowledge) as an intrinsic feature of the process that the belief is about. It is reasonable for me to believe, say, pretty strongly that Rock Roi will win the Cup. Forgetting what makes it reasonable for me to believe this, and, more important, forgetting that for other people, with other information, it may be reasonable to believe this less strongly or more strongly, I mistake this reasonableness-for-me for some objective tendency-for-Rock-Roi-to-win considered as an intrinsic feature of the situation in which the race is about to be run.[26] Whereas the logical relation is called probabilification because it is indeed the making something probable-in-the-central-sense, and a ratio among possibilities or a frequency may be called a probability by a sort of courtesy, because it is part of the evidence for a probability-in-the-central-sense, an objective chance is called a probability (or a chance) because it is a probability-in-the-central-sense detached from its moorings. Logical relation theory, classical and range theory, and frequency theory all shift probability terminology back from the conclusion to something in what supports the conclusion; objective chance theory shifts probability terminology forward to the actual event or process that this conclusion describes. On the other hand, we could assimilate the objective chance shift to the pattern of the classical and frequency ones. If there were attached, to a certain initial stage of any individual process, an objective chance of $x\%$ for such-and-such to occur, knowing this would indeed constitute an inconclusive reason for believing (once the initial stage of the individual process in question had been observed) that such-and-such would follow, and the larger x was, the better would this reason be. I shall consider (in § 7) *why* this is so.

If we take the informal concept of probability as central, then, it is easy to understand the shifts of meaning by which we first concentrate on what I have called simple probability—the good reasons end of the informal concept—and then move by way of it to the various other concepts and uses of probability language.

§ 7. *Valid and Invalid Transitions*

This multiplicity of distinct but connected probability concepts complicates the logic of probability statements. Some sorts of

[26] Cf. 'A Refutation of Morals', as in n. 9, p. 167, above.

move are valid for one concept but not for another, and some moves between concepts are valid while others are not.

In the objective chance sense of 'probability', if one person says that the probability at the start of the race that Rock Roi will win is 0·5, and another that it is 0·2, they cannot both be right. But if they say these things in the simple probability sense, they can both be right; if they are in possession of different information, it may be reasonable for one to bet at evens and the other at four to one against. Yet there is still a kind of conflict between these statements: to adhere to either is implicitly to condemn the other person's information as either defective or mistaken. Even this conflict disappears if we go on to a logical relation sense: 'This information probabilifies this conclusion to degree 0·5' and 'That information probabilifies the same conclusion to degree 0·2' can coexist with no strain at all. The conflict would disappear in another way if the probability statements were taken in a purely subjective sense—which is, perhaps, one reason why the logical relation view has been thought to be close to subjectivism.

Although the objective chance of a certain outcome, if there is such a thing, must have a definite value at a specific time, its value can change over time. If there were objective chances about sporting events, it might well be correct to say (as the commentators appeared to be saying) that just before the third test against India began England's chance of winning was about 0·6; at the end of the first day's play it was about 0·8; at the end of the third day's play 0·6 again; at the end of the fourth day's play about 0·3; and at the start of the last over perhaps 0·01. But it seems that a radium atom's chance of decaying in the next five minutes never alters until it actually blows up.

The objective chance that Tom will marry Susan is necessarily the same as the objective chance (at the same time) that Susan will marry Tom. The simple probabilities of these are also necessarily equal, from any single point of view. But things change as soon as we substitute propositional functions for propositions. The probability that a man will marry a woman need not be the same as the probability that a woman will marry a man. If this is interpreted as a frequency, it is clear that the frequency of marrying among men may be different from the frequency of marrying among women: ruling out such complications as polygamy, we shall necessarily have twice the percentage of married men among the men as of married women among the women in a community in which there are twice as many women as men. The same holds if a propositional function probability statement is interpreted as meaning 'If something is known simply as a man, it is reasonable to give to the statement that it is married the degree of belief measured as $x\%$'.

A simple fallacy in an attempt to prove a 'rule of succession' illustrates how we may be tempted to move improperly from one probability concept to another. Suppose that I have a pile of a hundred similar plates. I drop ninety-nine of them in turn from a height of three feet on to the floor, and they all break. I am about to drop the hundredth in the same way. If it breaks, the probability of a dropped plate from this set breaking will be 100%; if it doesn't break, it will be 99%; so this probability is at least 99%. So it is reasonable for me to be at least 99% sure that this last plate will break when it is dropped. And, generalizing, if any event has happened without fail on n occasions, the chance of its happening on the $n+1$th occasion is at least $n/(n+1)$.

But of course this is wrong. And what is wrong is that we have gone straight from a probability interpreted as a frequency in a finite class to a simple probability about an individual member of that class. I shall indeed argue that such a move is valid provided that nothing is known about this individual, except its being a member of this class, that is relevant to the question whether it has the characteristic in question. But this restriction is not being observed. The hundredth plate-dropping is not just any old member of the series within which the frequency is at least 99%: it is a very special member, the only one not known to result in breakage. We cannot validly go from a frequency in a finite class to a simple probability about such a distinctive member. But there are two things that make this fallacy just a little bit tempting. One is that what is distinctive about this member of the class is merely negative, it only *lacks* a known description to which the others conform, so that in a way nothing (positive) that is relevant and distinctive is known about it. The other is that the conclusion is intrinsically plausible: we believe that inductive reasoning is reasonable. Perhaps it is; but this method of proving that it is will not do.

All of these are elementary points; a more difficult question concerns the validity of detachment, or *modus ponens*, with probability statements and, what is connected with this, the validity of the proportional syllogism. Can we argue 'If P, probably Q; and P; therefore probably Q' either with or without a numerical value for the probability? And is the proportional syllogism as set out in § 2 above valid?

If the premiss 'If P, probably Q' is a genuine conditional (material or non-material) with the probability statement, of whatever sort, occurring wholly within the consequent, then of course detachment is valid without restriction. But the relational formula '$P(Q,P) = x\%$' can be read as 'The probability of Q, given P, is $x\%$' or 'The probability of Q, if P, is $x\%$' and this can slide into 'If P, probably (to $x\%$) Q'. This is not a simple conditional, though it looks like one: the probability operator governs the 'if', and not vice versa, so that detachment is not automatically in order here. Since what is being said is that P probabilifies Q to degree $x\%$, it is not immediately clear what follows if we combine this with the information that P. One thing that cannot follow is that the probability of Q is $x\%$ in the same sense of probability that we had in the premiss. Since that was a logical relation probability, this conclusion would not even be well formed. But perhaps we can read the conclusion as asserting a simple probability? This conclusion would be well formed, but it cannot follow without restriction, since we may well have the following situation:

(i) P probabilifies Q to degree $x\%$
(ii) R probabilifies Q to degree $y\%$
(iii) $x \neq y$
(iv) P is known to be true
(v) R is known to be true.

If detachment as suggested above were valid without restriction, the simple probability of Q would have the incompatible values $x\%$ and $y\%$. The restriction needed is Carnap's principle of total evidence.[27] That is to say, 'The simple probability of Q is $x\%$' follows not from 'The probability of Q, if P, is $x\%$' and 'P' on their own, but only from these provided that the point of

[27] *The Philosophy of Rudolph Carnap*, ed. P. A. Schilpp, p. 972.

view from which there is to be this simple probability includes no other information that bears on the question whether *Q*.

There is nothing mysterious or *ad hoc* about this requirement.[28] Once we see that the object of the exercise is to arrive at a simple probability, at a decision whether it is reasonable for someone to believe that *Q*, or how firmly he is entitled to believe this, it is obvious that whatever relevant information he has must be taken into account. Of course it would not be reasonable for someone to believe that *Q* merely because he has one piece of information which in itself strongly supports *Q*, if he also has some other piece of information that tells against *Q*. This requirement looks arbitrary only if we do not realize that we are moving from a relational probability—and of course there can be any number of different relational probabilities connecting *Q* as conclusion with various bits of evidence—to a simple probability, to what it is reasonable for someone to believe, which cannot be multiple in this way, and which must depend on all, not merely on some, of the information available to him. We are not preferring, rather arbitrarily, one relational probability to another, but going to a probability of a different sort.

Another way of looking at the matter is suggested, however, by the remark in § 2 above that a simple probability statement is *close* to a relational probability statement in which the evidence is referred to as 'my total present state of knowledge and ignorance', though not synonymous with this. We can make this clearer by introducing two speakers. *A* will be justified in saying 'Probably (to degree x%) *Q*' in exactly those circumstances in which *B* can say correctly '*A*'s total relevant information probabilifies *Q* to degree x%'.

But instead of thus *almost* reducing simple to relational probability, we could look at it the other way round. What *is* support or quasi-entailment or probabilification? We decided that it must have a formal basis. But it is not the form as such that makes it support, etc., but precisely that it is a form such that if we knew *P* and nothing else that bore upon the question whether *Q*, we should be justified in giving *Q* the degree of belief measured by x%. In other words, we could define probabilification so that detachment, subject to the stated restriction, was analytically valid.

[28] Contrary to what is said in A. J. Ayer, *The Concept of a Person*, pp. 191–4.

It is hard to say which of these two approaches is preferable; I shall come back to this question in § 9 below. What matters now is that whichever way round we look at it, detachment from relational to simple probability, subject to the restriction stated, is analytically sound. But, as usual, this means that the serious problems break out elsewhere: the question now will be whether this or that formal relationship does constitute probabilification as it is now being understood.

This brings us back, first, to the proportional syllogism. In at least some elementary cases this is 'intuitively' convincing. If I know that there are ten balls in a bag, nine of them black and one white but indistinguishable to touch, and I put my hand into the bag, stir them around, and then grasp one, am I not entitled to be fairly sure that the one I have is black? If my life depended on saying correctly what colour this ball was before I drew it out and looked at it, would it not be silly, or at least suicidal, to say 'white'? And would not the case for believing that the one I have grasped is black have been stronger if there had been ninety-nine black balls and only one white one? But why is it so plausible to say this sort of thing? Surely there is some, perhaps only half-formulated, reasoning behind the alleged 'intuition', and this reasoning makes use of the principle of indifference. I don't know which of the ten balls it is that I have grasped; it might be black number 1, it might be black number 2, and so on. Since I have no reason for giving a higher degree of belief to one rather than any other of these possibilities, it would be arbitrary to do so; to avoid arbitrariness, I must believe or disbelieve them equally, that is, I must give the same degree of partial belief to the possibility that I have black number 1, that I have black number 2, . . . and that I have the white one. So I must believe more strongly that I have *some one* black ball than that I have the white, and the more black balls there are, the greater the difference. The principle of indifference seems not to be analytic, but to be sound as an avoidance of arbitrariness; and once it is accepted, the validity of the proportional syllogism follows, in two ways. First, the stated information plus my lack of any other relevant information yields the appropriate simple probability for the conclusion 'This ball is black'; and, secondly, the information that nine out of the ten balls are black and this is one of them in itself probabilifies the conclusion that this is black to degree 0·9. All the stock objections to the principle of indifference, that it is not applicable to many sorts of problem, that if carelessly used it produces absurd results, that it is too slight a foundation for probability theory, that we want to pay attention to frequencies and perhaps other sorts of empirical data, do not affect its basic soundness or its applicability to at least some simple sorts of problem.

I spoke in § 2 about making a leap from using the proportional syllogism about a class of actual things or events of which x% are B to using it with regard to possibilities. But the present discussion shows that we are dealing with possibilities all the time. The balls in the bag are concrete or actual enough, but what I am thinking about is the *possibility* that the one I have is black number 1, and so on. Once we see how the validity of the proportional syllogism rests on the reasonableness of the principle of indifference as a way of handling possibilities-under-consideration, it seems that there could be no good reason for confining its application to concrete cases.

In particular, we can use it to derive a simple probability from a limiting frequency. Suppose that I know that if this penny as it now is were tossed in a certain manner indefinitely, the result heads would occur with a limiting frequency of, say, 0·6, but with heads and tails occurring in a random order. Then I must regard any particular toss (in that manner) of this penny as it now is—e.g. the very next toss—as a member of such a hypothetical sequence. But I do not know which member of the sequence it is, not even whether it is an early member or a late one, since this hypothetical sequence can be supposed to stretch back into the past as well as forward into the future. So, by the principle of indifference, I must be equally prepared to identify this toss with each member of the sequence; distributing belief equally among these possibilities, I am committed to giving the degree of belief 0·6 to the possibility that this toss is one or other of those that have come up heads. Of course, I cannot in these circumstances apply the principle of indifference to the two possibilities 'heads' and 'tails' themselves, to yield the simple probability of 0·5 for heads; for I do have the other relevant information about the limiting frequency. It is worth stressing that while frequency theorists have commonly attacked the principle of indifference, it is only by this principle that a frequency, either in a finite class or a limiting frequency, is linked to a simple probability.

It does, indeed, seem that we could justify the transition from a limiting frequency to a simple probability in another way, using Bernoulli's Central Limit Theorem (discussed in § 8 below) as part of a *reductio ad absurdum* argument. Suppose that I have accepted as well confirmed the hypothesis that there is a limiting frequency 0·6 of heads in tosses of this penny, and that the sequence of these tosses is a collective. Then suppose that I give to the possibility that this penny will fall heads at the next toss any degree of belief *other* than 0·6—say 0·4. To avoid arbitrariness, I must then give the same degree of belief to the possibility that it will fall heads at the second toss, and similarly for the third toss, and so on indefinitely. Since the sequence was assumed to be a collective, it is a Bernoulli sequence, and hence by Bernoulli's Theorem I must give a very high degree of belief to the possibility that the frequency of heads in any long series of tosses is close to 0·4, and that the limiting frequency is 0·4. But this would be inconsistent with my having accepted the hypothesis that the limiting frequency is 0·6. That is, I cannot coherently ascribe one value to the limiting frequency for heads, say, in a collective and any other value to the simple probability of heads at the next toss which is a member of that collective, about which nothing special and relevant is known. So I can argue from the limiting frequency to the corresponding simple probability at the next toss.

This seems correct. But it is not essentially different from a direct appeal to the principle of indifference. This argument too is not deductively valid, but turns upon the avoidance of the arbitrariness of assigning one simple probability value to heads at the next toss, a different value to heads at the toss after next, and so on. This avoidance of arbitrariness is the central notion in the principle of indifference, as was noted in § 2 above.

One reason for hostility to this principle is what I might call the Puritanical objection that its use is an avoidance of honest toil. Why bother to investigate frequencies, etc., if we can get, perhaps, quite high probabilities *a priori*?[29] Some have tried to

[29] Cf. K. R. Popper, *Logic of Scientific Discovery*, pp. 407–8. Popper uses the 'paradox of ideal evidence' to refute what he calls the subjective theory of probability,

tighten up the principle to prevent such misuse. Kneale, for example, insists that 'we are entitled to treat alternatives as equiprobable if, but only if, we know that the available evidence does not provide a reason for preferring one to the other', and *not* if we merely 'do not know that the available evidence provides (such) a reason'. Should simple probability be treated as relative to the information that is *available* rather than to the information that someone actually has? Mellor goes further, saying 'We are entitled to treat alternatives as equiprobable if, but only if, we know that the available evidence provides a reason for not preferring any one to any other'.[30]

However, Mellor's formulation is too restrictive. He is endeavouring to make the principle analytic; but I think we are justified in using a principle which falls short of analyticity but crystallizes what I have called the avoidance of arbitrariness. Admittedly in following such a principle we can go wrong.

by which he means a view that identifies probability exclusively either with what I have called simple probability or with probabilification (as in the logical relation account). On these views, the probability that this penny will fall heads at the next toss is 0·5 *both* before the penny has been measured, tested, etc., *and* after it has been thoroughly checked for lack of bias both by direct physical measurement and by its showing a frequency for heads of 0·5 in a long series of tosses. On what Popper calls the 'subjective theory', the measure of rational belief is not altered by the addition of a great deal of further information which is in any ordinary sense highly relevant. The usual comment is that there is another feature, the 'weight of evidence', over and above the measure of simple or relational probability. Popper argues that to admit this is to undermine the 'subjective theory', which is wedded to a linear or unidimensional treatment of the rationality of belief. We want to say that on 'zero' evidence and on 'ideal' evidence alike, the probability that the penny will fall heads at the next toss is 0·5, but that the probability that the probability of this is 0·5 is much higher on ideal than on zero evidence. Now it is true that in the latter statement, the word 'probability' cannot stand for simple (or logical relational) probability both times. The simple or logical relational probability of heads-at-the-next-toss just *is* 0·5 on either evidence, so the simple or logical probability that the simple or logical probability is 0·5 must be 1. What is true is that the simple or logical probability on 'ideal' evidence that the limiting frequency for heads in tosses of this penny is 0·5, or perhaps that this penny has an objective chance of 0·5 of falling heads at each throw, including the next, is much higher than it was on 'zero' evidence.

This shows, indeed, that we cannot deal adequately with this sort of example in terms of the concepts of simple or relational probability alone: we need either the concept of limiting frequency or that of objective chance or propensity as well. But it does not show that the concepts of simple or relational probability are incoherent, or that there is anything wrong with the use of the principle of indifference in connection with them: only that there are other aspects of probability *as well.*

[30] W. Kneale, *Probability and Induction*, pp. 172–3; D. H. Mellor, *The Matter of Chance*, p. 138.

A simple probability indicates what *in our present state of knowledge and ignorance* it is reasonable to believe, but this in no way denies that there may be better states, ones containing more knowledge and less ignorance. The simple probability relative to 'ideal' evidence may indeed turn out to have the same value as the simple probability relative to minimal evidence; but it may not. If one has only minimal evidence, and additional information might raise or lower the simple probability value for heads at the next toss, and such information—e.g. whether the penny has shown a bias in a series of previous tosses—is available for the asking, then if one is thinking of betting a large sum on the result of the next toss it would be sensible to seek this information first. If we learn that the penny has shown no bias, we are indeed no better placed than we were for betting on the next toss; but if we learn that it has shown a bias, we are better placed. Similarly the man who uses the principle of indifference in Mellor's form has more information than he would have needed in order to use it in Kneale's form; but the extra information might have led to a different simple probability distribution. If we have only the Kneale-type information, it will be reasonable to treat the alternatives as equiprobable until we learn more, and perhaps also when we learn more. But it would be silly to treat the principle of indifference as a discouragement from learning more: its function is merely to indicate what it is reasonable to believe in the light of whatever information we now have.

What it is, at any moment, reasonable for someone to believe is relative to the information that he then has. But of course if he could easily get further relevant information, and the problem is of some importance to him, it will be sensible for him to get that further information first, *before* he resorts to the principle of indifference, even though he will still in the end have to resort to it in order to derive a simple probability. A judgement based on a greater quantity of relevant information is a better judgement. And if we are asked *how* it is better, we can say without blushing that it is more likely to be right. Admittedly in the particular case where a penny is in fact unbiased, the man who bases his opinion as to whether it will fall heads at the next toss on the principle of indifference applied directly to the alternatives 'heads' and 'tails' will be no less likely to be right than the man who, by a long series of trials, first confirms the hypothesis that the limiting frequency for heads is $\frac{1}{2}$ and then applies the principle of indifference (or what comes to the same thing, a proportional syllogism). But this *is* a particular case. If we start off not knowing whether the penny is biased or not, it is clearly reasonable to say that the man who experiments thoroughly first is more likely to be right even about a single later result.

As we saw in § 2, once we have the range method of representation, including some assignment of measures, detachment, with the above-mentioned restriction, holds automatically. If $x\%$ is the ratio of the measure of *h* & *e* to the measure of *e* in some larger realm of possibilities, this ratio still holds when the universe of relevant possibilities is cut down just to *e*. The same holds for the first sort of weak probabilification, as long as we go on taking *f* for granted. With the second sort of weak probabilification, it is clear that in the conditions stated in § 2, the ratio of the measure assigned to *h* & *e* to that assigned to *e* must be equal to the measure assigned to *g*, so when we cut down the universe of relevant possibilities just to *e*, we must still give *h* the same simple probability that we gave to *g*. But we can hardly use this style of representation and its consequences to justify the sorts of detachment we want; we must rather argue independently that our procedures are reasonable, and hence that this style of representation is appropriate. The essential procedure, translated out of the range terminology, is that of holding *relative* degrees of belief constant when we exclude possibilities that we were originally leaving open, and it does seem that consistency demands this.

Another important transition is from an objective chance to an informal or simple probability. Someone who was sceptical about objective chance might regard this move as analytically valid: since, for him, objective chance would be a fiction to be explained as a projection of a degree of belief whose being justified would constitute a simple probability, an objective chance statement would come as near as such a fiction could come to being correct just when the corresponding simple probability statement held. But if we take objective chance seriously we must argue in another way: we must say that since (as we saw in § 4) an objective chance is something which would produce a corresponding limiting frequency, and since we could derive the corresponding simple probability from this limiting frequency, we can also derive it from the objective chance itself. Since we really know nothing else about objective chances except that they are such as to generate frequencies, I think that this is the *only* legitimate way of making this transition.

We can arrive, then, at a simple probability about the same matter in a number of different ways. For this very reason, it will not be in general valid to argue back from a simple probability to any one of the things that might have generated it. For example, consider the simple probability statement (about a die that has just been thrown, but where the result of the throw has not yet been inspected by or reported to the person for whom this statement is to express a reasonable degree of belief) 'The probability that it has fallen 5 up is $\frac{1}{6}$'. This statement would be justified if this person was in any one of the following states of incomplete knowledge:

(i) He knows nothing relevant except that the die has six sides marked '1' to '6', and that it cannot help resting with one side uppermost. Among other things, he doesn't know that it isn't biased.

(ii) He has been told that the throw in question is one of a series of six hundred throws in which the six sides come up with approximately equal frequency, but he has not been told the result of this particular throw.

(iii) He has been told that in a long series of throws the frequency with which side 5 came up looked as if it was settling down to a limiting frequency of $\frac{1}{6}$, but the result of this particular throw has, as I said, not been observed.

(iv) He has made observations which confirm the hypothesis that the die, thrown in a certain way, behaves indeterministically, with an objective chance of $\frac{1}{6}$ for each of the six sides to come up, and he has observed that the die was thrown in the appropriate way this time.

I have argued that in each of these four cases he has to appeal, implicitly at least, to the principle of indifference in order to arrive at a simple probability, but he applies this principle to a different set of data in the four cases—in (i) to the six alternatives, 'the die fell 1 up this time', etc., in (ii) to the six hundred alternatives identifying this throw with each of the six hundred throws in turn, in (iii) to an indefinite series of identifications of this throw with the various members of a hypothetical sequence of throws, and in (iv) to the same series as in (iii), since, as I have argued, objective chance connects with simple probability via limiting frequency.

Since he might properly have arrived at the simple probability judgement (that the chance that the die fell 5 up this time is $\frac{1}{6}$) in any one of these four ways at least, we cannot argue from the fact that this informal judgement is justified to any one of these bodies of incomplete information. We cannot conclude that nothing is known about the die except that it has the specified six sides; nor that there has been any particular frequency in a finite series of throws; nor that its throws form a collective with a limiting frequency of $\frac{1}{6}$ for 5 up; nor that at the start of this throw there was an objective chance of $\frac{1}{6}$ that 5 would come up. Equally we cannot argue validly from any one of these bodies of incomplete information to any other, with one exception. If at every throw there is an objective chance of $\frac{1}{6}$ for 5 up, then there must also be an equal limiting frequency in the hypothetical sequence of throws: this follows from the meaning of 'objective chance'. But the converse does not hold, nor would any of the other conceivable transitions. In particular, it is blatantly fallacious to argue from the minimal knowledge in (i) to the conclusion that there has been or will be any particular frequency either in any finite series of throws or in an indefinitely extended one. It is, I think, this pattern of inference in particular that has given the principle of indifference a bad name and has invited the accusation that to use it is to attempt to get knowledge out of ignorance. We cannot, certainly, get knowledge of a frequency or of a frequency-generating chance out of the fact that *we* have no reason to expect one outcome rather than another. But to say this is to condemn a misuse of the principle, not the principle itself.

§ 8. *Bernoulli and Bayes*[31]

The Law of Large Numbers established by Bernoulli's Central Limit Theorem seems to provide important links between

[31] This section is both heavily indebted to and in part critical of J. R. Lucas, *The Concept of Probability*, chs. 5 and 8.

various probability concepts. But if it is to be used correctly, two things need to be made clear. One is the mathematical content of the theorem, which is closely related to the way in which it is proved; the other is the interpretation, or perhaps interpretations, that can properly be put upon it. The content is explained very carefully by Lucas, whose account I follow. The theorem applies only to Bernoulli sequences, that is, to cases where we have a set of independent propositions which are all instances of the same propositional function and all have the same probability, e.g. the propositions that this coin will fall heads at the first toss, at the second toss, and so on. If the probability of heads at each toss is, say, α, what the theorem says is this: 'If you specify how close to α the frequency (in a finite sequence of tosses) must be to be acceptable, and how close to 1 you want the probability of our having an acceptable frequency to be, I can find an N, so that in any Bernoulli sequence with N or more instances, the probability of the frequency being an acceptable one is within the degree of closeness to 1 that you have desired.'[32] More simply but not quite so accurately: 'If the probability of heads at each toss is α, the probability that the frequency of heads in a series of N tosses will be close to α can be made close to 1 by making N large enough.'

I want to consider, first, what are legitimate interpretations of this theorem, and secondly, whether and in what circumstances it can be used to link observed frequencies with probabilities.

The theorem can be legitimately interpreted as a statement about objective chances: if there is an objective chance a of heads at each toss, then there is a very high objective chance that the frequency of heads in a long finite series of tosses will be close to α. E.g. if each toss of our penny were an indeterministic device tending, to degree 0·5, to produce heads, a series of, say, a thousand tosses would be another indeterministic device tending, to a very high degree, to produce a set of results in which there were between 480 and 520 heads.

Again, it can be legitimately interpreted as a statement about limiting frequencies: if in a random series (in von Mises's sense) of tosses there is a limiting frequency α of heads, then this series

[32] Lucas, op. cit., p. 83.

can be regarded as a series of sets of N tosses, and in *this* series the limiting frequency of N-membered-sets-with-a-proportion-of-heads-that-is-close-to-α will be close to 1 if N is large.

Also, in view of the moves already justified from objective chance and limiting frequency to simple probability, and between simple probability and probabilification, the following further interpretations are also valid.

If it is known that there is either an objective chance α of heads at each toss of this penny, or a limiting frequency α of heads among tosses of this penny, then it is simply probable to a high degree that the frequency of heads in a long finite series of tosses will be close to α, provided that nothing else is known that is relevant to the results of these tosses, i.e. especially provided that these results are not yet known themselves.

'There is an objective chance α of heads at each toss' probabilifies highly 'The frequency of heads in this long series of tosses is close to α'; and so does 'There is a limiting frequency α of heads among tosses of this penny'.

But can the theorem be legitimately interpreted as a statement about simple probabilities which are not themselves derived from objective chances or limiting frequencies, but from sheer ignorance by way of the principle of indifference? If it is reasonable for me to expect heads, at any one toss of this coin, no more and no less than I expect tails, because all I know is that it must fall one way or the other and I have no reason to prefer either, does it follow that it is reasonable for me to be very confident that in a series of a thousand tosses there will be between 480 and 520 heads?

Prima facie this seems implausible. Our natural suspicions about it are reinforced if we consider another slightly different situation. Suppose that I know that this penny is strongly biased one way or the other, but I don't know which—I have been reliably informed, say, that the limiting frequency for heads is *either* 40% *or* 60%, but I have not been told which. Then it is still reasonable for me to expect heads, at each individual toss, no more and no less than I expect tails, i.e. to assign a simple probability of $\frac{1}{2}$ to heads at each toss. But it would be utterly unreasonable of me to be very confident, in these circumstances, that in a series of a thousand tosses there will be between 480 and 520 heads. If anyone thinks this would be reasonable, let him consider a yet more extreme case: I am told that the penny which is to be tossed is either a double-headed or a double-tailed one, but I am not told which. The simple probability for heads, based on my ignorance and the principle of indifference, is still $\frac{1}{2}$; but I can now deduce validly that the frequency of heads in a thousand tosses will be either 1 or 0, and certainly not in the neighbourhood of $\frac{1}{2}$.

What, then, has gone wrong? If reasonable degree of belief can be measured as a probability and should obey the standard calculus, why does Bernoulli's Theorem, which is provable within that calculus, not apply to it? In the last case, the answer is easy: the propositions 'this coin will fall heads at the first toss', 'this coin will fall heads at the second toss' and so on are not *independent*: the information given entails that either all of them are true or all of them are false, so the argument in the proof of Bernoulli's Theorem simply does not apply. The same is true, in a less blatant way, in the second last case. If I know that the penny is biased one way or the other, the statements 'it will fall heads at the first toss' and 'it will fall heads at the second toss' cannot be regarded as having independent probabilities of $\frac{1}{2}$. Whichever way the first toss goes it will, so to speak, tend to carry the second toss with it. Once again we do not have a Bernoulli sequence *with a probability of $\frac{1}{2}$ for heads each time*, though—and this is the confusing factor—there will now be a Bernoulli sequence with *some other* probability for heads each time, namely *either* 0·4 *or* 0·6; in relation to this other frequency, the statements asserting heads at the first toss, at the second toss, and so on will be independent.

If we now go back to the first case, where I know nothing about the coin except that it will, at each toss, either fall heads or fall tails, and do not know either that it isn't biased or that it is, I cannot validly assume that 'heads at the first toss', 'heads at the second toss', and so on all have *independently* the probability $\frac{1}{2}$. I must, by the principle of indifference, give each of these the probability $\frac{1}{2}$, but I cannot argue validly in the way required for the proof of the theorem because I must allow for the possibility that they may not be independent.

In other words, we cannot combine this way of deriving the probability of $\frac{1}{2}$ for heads at each throw, from ignorance via the principle of indifference, with the treatment of the statements describing the results of tosses as a Bernoulli sequence. So our suspicions were justified: the theorem cannot be applied to simple probabilities based on sheer ignorance. This does not, of course, cast doubt on the principle of indifference itself or the probabilities reached by reliance on it; it shows only that they do not have the sort of independence required to allow the proof of Bernoulli's Theorem to go through.

Is there a legitimate interpretation of this theorem as a statement not about limiting frequencies but about frequencies in a finite class? That is, does the following formulation hold: If in a long finite sequence of tosses there is a frequency α of heads, the frequency of N-membered-sub-sequences-with-a-proportion-of-heads-close-to-α will be close to 1 if N is large? Whether this holds or not depends on how we understand the class of N-membered-sub-sequences in which we are hoping to find a very high proportion of ones that each have a proportion of heads close to α. If we take this as the class of all N-membered sub-sets of the main sequence, i.e. if we think of constructing sets with N tosses in each in all logically possible ways, then indeed the theorem holds. But if we think of, say, sequences of the first N consecutive tosses, the second N consecutive tosses, and so on, or of N-membered sub-sequences chosen by any other selection procedure, the theorem may not hold, again because we have not insisted on independence. And it seems that we cannot here insist on independence, because a finite sequence cannot be random in von Mises's sense. On the other hand, a finite sequence can be apparently random, it can be immune to all the more obvious methods of place-selection. If it is, then the theorem will still hold when applied, say, to sub-sequences each of which consists of N consecutive tosses; the proportion among these of ones that have each a proportion of heads close to α will still be close to 1 if N is large.

Given that the theorem has these legitimate interpretations, can it be used as part of an 'inverse' argument, to argue back from an observed frequency to a probability of some sort? We cannot, of course, simply *convert* the theorem: we cannot say straight away that 'The frequency of heads in this long series of tosses is close to α' probabilifies highly 'There is an objective chance α of heads at each toss' or 'There is a limiting frequency α of heads among tosses of this penny'. 'P probabilifies Q' does not entail '*Q* probabilifies *P*', any more than 'Most *A*s are *B*' entails 'Most *B*s are *A*'. But we may be able to get something like this result indirectly, using, perhaps, an analogue of contraposition rather than of conversion, or the falsification rather than the verification of hypotheses. In fact, the result we might hope to get would be that 'The frequency of heads in this long series of tosses is α' probabilifies highly 'There is a limiting frequency close to α of heads among tosses of this penny' or 'There is an objective chance close to α of heads at each toss'.

Take the objective chance interpretation, for example. If the coin has at each toss a chance close to α of falling heads then if it is tossed a large but finite number of times it has a very high chance of generating a sequence of results in which the frequency of heads is close to α. Let us frame the hypothesis that there is an objective chance close to 0·6, say, for heads at each toss. Suppose that in 1,000 tosses the observed frequency is not close to 0·6, but is, say, 0·423. Or, to make the evidence stronger, suppose that the coin is tossed 100,000 times and falls heads just 42,157 times. Does either body of evidence falsify or disconfirm the hypothesis, and show that at any rate there is not an objective chance close to 0·6 of heads at each toss? Well, certainly this hypothesis is not *falsified.* A coin with an objective chance close to 0·6 of heads has a non-zero objective chance of producing 423 heads in 1,000 tosses and even of producing 42,157 heads in 100,000 tosses, and it may be this chance that has been realized this time. On the other hand, we should like to say that the hypothesis has been disconfirmed, but it is not clear why it has been.

There are two ways in which we might try to formulate the argument:

(*a*) The first runs thus:

1. If there is an objective chance close to 0·6 of heads at each toss, there is a very small objective chance of just 423 heads in 1,000 tosses.
2. But 423 heads in 1,000 tosses have occurred.
3. Therefore, there is not an objective chance close to 0·6 of heads at each toss.

If we argue in this style, we are treating the occurrence of something as being like a falsification of the statement that there is a very small objective chance of its occurring, and then using *modus tollendo tollens*.

(*b*) The other formulation runs:

1. If there is an objective chance close to 0·6 of heads at each toss, there is a very small objective chance of just 423 heads in 1,000 tosses; that is, there is a very big objective chance of there not being just 423 heads in 1,000 tosses.
2. 'There is an objective chance close to 0·6 of heads at each toss' probabilifies highly 'There are not just 423 heads in the first 1,000 tosses'.
3. 'There are just 423 heads in the first 1,000 tosses' probabilifies highly 'There is not an objective chance close to 0·6 of heads at each toss'.
4. There are just 423 heads in the first 1,000 tosses.
5. Therefore, there is not an objective chance close to 0·6 of heads at each toss.

If we argue in this style we have relied on the following kinds of move: in going from 1 to 2, we have derived a probabilification from a high objective chance. In going from 2 to 3 we have contraposed a probabilification statement. In going from 3 and 4 together to 5 we have used detachment. Now this detachment is of a kind I have already defended; it will be valid provided that we have no other evidence that bears upon the objective chance of heads at each toss. The move from 1 to 2 is all right, because the objective chance statement allows us to derive a simple probability, and for reasons given above this will yield a probabilification. The critical step is that from 2 to 3.

Now Lucas says that in probability theory the Law of Contraposition and *modus tollens* do not apply:[33] that is, neither style of argument is correct. Nevertheless, he understands our reasoning in this sort of case in terms of model (*a*). He says 'Our strategy is to move by means of Bernoulli's Theorem from the normal ranges of probabilities where *modus tollens* does not work at all to the special extremes where it almost does.'[34] He backs this up with an argument that truth should be identified not just with the point 1 on the probability scale, but with the *neighbourhood* of 1, and falsehood likewise with the *neighbourhood* of 0.

[33] Op. cit., p. 126. [34] Op. cit., p. 127.

This in turn is defended by looking at the problem dialectically, in terms of a debate between two speakers. The man who is using Bernoulli's Theorem asks his opponent to say how extreme a coincidence must be, on his hypothesis, to make him reject the hypothesis that makes it a coincidence; he then carries out a large enough number of trials to ensure that, by the theorem, the observed result would be more of a coincidence than his opponent is prepared to tolerate. To be consistent, the opponent must now reject his hypothesis. Alternatively, if the opponent refuses to set any limits, if he makes it clear that he will go on for ever postulating more and more extreme coincidences in order to save his original hypothesis, he is not worth arguing with.[35]

To say this, however, is to bludgeon one's way out of a difficulty. This last argument mixes up two points. One is that you are entitled to ask your opponent how improbable something has to be to make him reject it, and to hold him to his bargain if he makes one, and to decline to argue with him if he will not make one. But you are so entitled just because what is (simply) improbable is something that it is reasonable to disbelieve. This correct point is mixed up with the thesis that you are entitled to ask your opponent how coincidental something has to be to make him reject whatever makes it a coincidence; this is quite another matter, it has no prima facie plausibility, and it would become plausible only if it could be assimilated to the first point by its being shown that whatever makes something else coincidental is itself improbable—but this is the very inversion that we are trying to justify. Until it is independently justified, the dialectical challenge is not in order.

Nor will it do to identify truth and falsehood with neighbourhoods. If we *could* identify falsehood with a very low probability, we could use argument (*a*) above to disprove, not merely to disconfirm, the hypothesis that the objective chance of heads at each toss is close to 0·6. On this hypothesis the statement 'There are just 42,157 heads in 100,000 tosses' is extremely improbable; if this improbability comes within the neighbourhood of 0 it is to count as false within the scope of this hypothesis. But since it is observed to be true, this hypothesis must be false. Of course we can't say this, as Lucas admits. Since we could

[35] Op. cit., pp. 84–94.

say it if we identified falsehood with the neighbourhood of 0 on the probability scale, this identification itself must be rejected. (In fact it is a mistake to identify truth and falsehood even with the points 1 and 0 on the probability scale: probability values lie between certainty and certainty-that-not; truth and falsehood are in a different league. But this is a different issue, and it would be for most purposes pedantic to press it; it matters here, however, because the initial confusion of truth with certainty and so with a probability value 1 facilitates the further error of identifying truth with the neighbourhood of 1 and falsehood with the neighbourhood of 0.)

Lucas has been driven to these desperate measures by the belief that we ought to be able to use Bernoulli's Theorem somehow in an inverse argument. Argument (*a*), at least as Lucas develops it, fails. But what about argument (*b*)? Lucas would reject this, because, as we saw, he says that in probability theory contraposition does not hold. But the example he gives is of the not holding of a contraposition of a propositional function, that we cannot argue from something of the form 'The probability that a *G* is an *F* is *x*%' to the corresponding statement of the form 'The probability that a non-*F* is a non-*G* is *x*%'.[36] Certainly we cannot, because this changes the reference class, as does the fallacious move from the probability of a man marrying a woman to the probability of a woman marrying a man. But in argument (*b*) we use not this form, but a move from '*P* highly probabilifies *Q*' to 'not-*Q* highly probabilifies not-*P*' where *P* and *Q* are not propositional functions but propositions. So let us look further at this contraposition: it will be easiest to use the range representation and a diagram.

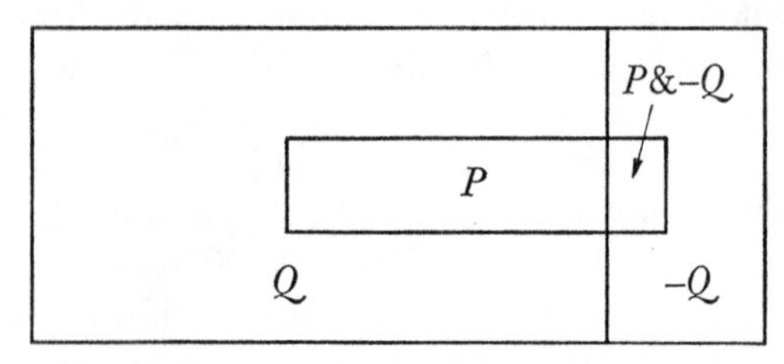

The outer rectangle represents the total range of possibilities, the left-hand division of it those in which *Q* holds, the right-

[36] Op. cit., p. 127.

hand one those in which not-Q holds. The inner rectangle encloses all the possibilities in which P holds. Then to say that P highly probabilifies Q is to say that Q holds in most of P's area, i.e. that the area for P & not-Q is only a small part of the area for P. To say that not-Q highly probabilifies not-P is to say that not-P holds in most of not-Q's area, i.e. that the area for P & not-Q is only a small part of the area for not-Q. Now these two are clearly not equivalent. If the area for not-Q were much smaller than for P, the area for P & not-Q could be a large part of the former but only a small part of the latter. So if the range representation is anything like appropriate, equivalent contraposition does not hold even for probabilification between propositions. Argument (*b*), therefore, fails at the vital step. But since argument (*a*) has also failed, and I can see no other way of using Bernoulli's Theorem in an inverse argument, I think the conclusion to be drawn is that this theorem cannot be used at all in this way.

But let us look at the range diagram again. It makes it clear that contraposition *would* hold (from 'P probabilifies Q' to 'not-Q probabilifies not-P') if the simple probability of not-Q were at least as great as that of P, i.e. in this case if the simple probability of 'There are just 423 heads in the first 1000 tosses' were at least as great as the simple probability of 'There is an objective chance close to 0·6 of heads at each toss'. Whether this is so or not, we can show that not-Q probabilifies not-P if we can show that the area for not-P & not-Q is much larger than that for P & not-Q, in other words that it is *relatively* easy to get not-Q from not-P, that is, that *alternative* hypotheses to P would give a much higher chance of getting just 423 heads in the first 1000 tosses. In fact, we *can* show this in this sort of case; but to do so we must go on from Bernoulli's Theorem to those of Bayes. Bernoulli's Theorem on its own will not sustain the inverse sort of argument.

I have discussed this point with reference to the objective chance interpretation. But if we started instead with a hypothesis about a limiting frequency or a finite class ratio we should still be blocked at the corresponding points when we tried to use either *modus ponens* or contraposition. Bernoulli's Theorem will not allow us to argue back, then, from an observed frequency in a finite series of trials either to an objective chance or to a limiting frequency or a frequency in a larger finite class of which the trials form a sub-class.

A second attempt to justify inverse probability reasoning is based on Bayes's Rule, which is formulated (in Lucas's symbolism) thus:[37]

$$\text{Prob}\,[H(gf)] = \frac{\text{Prob}\,[F(gh)] \times \text{Prob}\,[H(g)]}{\text{Prob}\,[F(g)]}.$$

That is, the probability that something which is both *G* and *F* should also be *H* is found by multiplying together the probability that something which is both *G* and *H* should be *F* and the probability that something which is *G* should be *H*, and dividing by the probability that something which is *G* should be *F*. This is a theorem of the standard probability calculus. But how is it to be interpreted and applied?

Lucas quotes from Sir Ronald Fisher the following illuminating example in genetics:[38]

In Mendelian theory there are black mice of two genetic kinds. Some, known as homozygotes (BB), when mated with brown yield exclusively black offspring; others, known as heterozygotes (Bb), while themselves also black, are expected to yield half black and half brown. The expectation from a mating between two heterozygotes is 1 homozygous black, to 2 heterozygotes, to 1 brown. A black mouse from such a mating has thus, prior to any test-mating in which it may be used, a known probability of $\frac{1}{3}$ of being homozygous, and of $\frac{2}{3}$ of being heterozygous. If, therefore, on testing with a brown mate it yields seven offspring, all being black, we have a situation perfectly analogous to that set out by Bayes in his proposition, and can develop the counterpart of his argument, as follows:

The prior chance of the mouse being homozygous is $\frac{1}{3}$; if it is homozygous the probability that the young shall be all black is unity; hence the probability of the compound event of a homozygote producing the test litter is the product of the two numbers, or $\frac{1}{3}$.

Similarly, the prior chance of it being heterozygous is $\frac{2}{3}$; if heterozygous the probability that the young shall be all black is $\frac{1}{2^7}$, or $\frac{1}{128}$; hence the probability of the compound event is the product $\frac{1}{192}$.

[37] Op. cit., p. 146. In this symbolism, which Lucas borrows from T. J. Smiley, upper-case letters stand for predicates and the corresponding lower-case letters for variables with correspondingly limited range. E.g., if '*G*' = 'is a mouse', then '*g*' is an individual variable ranging over mice.

[38] Lucas, op. cit., p. 147; R. A. Fisher, *Statistical Methods and Scientific Inference*, 2nd edition, pp. 18–20.

But, one of these compound events has occurred; hence the probability after testing that the mouse tested is homozygous is

$$\tfrac{1}{3} \div (\tfrac{1}{3} + \tfrac{1}{192}) = \tfrac{64}{65},$$

and the probability that it is heterozygous is

$$\tfrac{1}{192} \div (\tfrac{1}{3} + \tfrac{1}{192}) = \tfrac{1}{65}.$$

If, therefore, the experimenter knows that the animal under test is the offspring of two heterozygotes, as would be the case if both parents were known to be black, and a parent of each were known to be brown, or, if, both being black, the parents were known to have produced at least one brown offspring, cogent knowledge *a priori* would have been available, and the method of Bayes could properly be applied.

In discussing this, let us use the following predicate symbols, and the small letters for individual variables correspondingly:

G: is a black mouse, the offspring of two heterozygous parents, which has been mated with a brown mouse and produced exactly seven offspring.
F: has only black offspring
H_1: is homozygous
H_2: is heterozygous

The Mendelian theory determines the following probability values:

$$\begin{aligned} \text{Prob}\,[H_1(g)] &= \tfrac{1}{3} \\ \text{Prob}\,[H_2(g)] &= \tfrac{2}{3} \\ \text{Prob}\,[F(gh_1)] &= 1 \\ \text{Prob}\,[F(gh_2)] &= \tfrac{1}{128} \end{aligned}$$

To find Prob $[F(g)]$ we have to note that what is G must be either H_1 or H_2, and then use both the multiplication rule of the calculus for the probability of combined events, and the addition rule for the probability of an exclusive disjunction of events. That is

$$\begin{aligned} \text{Prob}\,[F(g)] &= \text{Prob}\,[FH_1(g)] + \text{Prob}\,[FH_2(g)] \\ &= \text{Prob}\,[H_1(g)] \times \text{Prob}\,[F(gh_1)] + \text{Prob}\,[H_2(g)] \times \\ &\quad\ \text{Prob}\,[F(gh_2)] \\ &= (\tfrac{1}{3} \times 1) + (\tfrac{2}{3} \times \tfrac{1}{128}) \\ &= \tfrac{1}{3} + \tfrac{1}{192}. \end{aligned}$$

Then substituting first H_1 and h_1 for H and h, and then H_2 and h_2 for H and h in Bayes's Rule we find in turn:

$$\text{Prob}\,[H_1(gf)] = \frac{1\times\frac{1}{3}}{\frac{1}{3}+\frac{1}{192}} = \frac{64}{65},$$

$$\text{Prob}\,[H_2(gf)] = \frac{\frac{1}{128}\times\frac{2}{3}}{\frac{1}{3}+\frac{1}{192}} = \frac{1}{65}.$$

But what does all this *mean*? I think the clearest interpretation is in terms of limiting frequencies, in a number of hypothetical indefinitely extended sequences.

Thus 'Prob $[H_1(g)] = \frac{1}{3}$' tells us that in such a series of mice that are G, the limiting frequency of homozygousness is $\frac{1}{3}$. (There is in fact an extra assumption here: all that Mendelian theory itself tells us is that this is the frequency among black mice that are the offspring of two heterozygous parents: we have to assume that the further characteristic of producing exactly seven offspring when mated with a brown mouse is independent of this and does not alter the frequency. But it might not be: it might be harder, or easier, to get seven offspring from a homozygous/brown mating than from a heterozygous/brown mating. But I shall continue to use the necessary assumption.)

With similar interpretations for the other probability values, we can trace the argument as follows.

We start with a hypothetical indefinitely extended series of mice that are G. Among these, the limiting frequencies are $\frac{1}{3}$ for being homozygous, $\frac{2}{3}$ for being heterozygous. Among the homozygous ones, the frequency of having only black offspring is 1, among the heterozygous ones, the frequency of having only black offspring (in the assumed family of 7) is $\frac{1}{128}$. So the frequency of being homozygous, and also having only black offspring is $\frac{1}{3}\times 1$ (i.e. $\frac{1}{3}$), while the frequency of being heterozygous and also having only black offspring is $\frac{2}{3}\times\frac{1}{128}$ (i.e. $\frac{1}{192}$). Now we select from the original series a sub-series containing all and only those that have only black offspring. These constitute a limiting proportion amounting to $\frac{1}{3}+\frac{1}{192}$ of the original series. Since the homozygous ones among them formed a limiting proportion of $\frac{1}{3}$ of the original series, these form a limiting proportion $\frac{1}{3}/(\frac{1}{3}+\frac{1}{192})$ (i.e. $\frac{64}{65}$) of the sub-series.

Similarly, since the heterozygous ones among them formed a limiting proportion of $\frac{1}{192}$ of the original series, these form a limiting proportion $\frac{1}{192}/(\frac{1}{3}+\frac{1}{192})$ (i.e. $\frac{1}{65}$) of the sub series.

Thus the final probability values, Prob $[H_1(gf)] = \frac{64}{65}$ and Prob $[H_2(gf)] = \frac{1}{65}$ can be interpreted as the limiting frequencies of homozygousness and heterozygousness respectively in an indefinitely extended series of mice that satisfy condition F as well as condition G.

An alternative but closely parallel interpretation would use a large finite class of mice that are G and the sub-class of these that are F. Provided that the numbers were large enough to display approximately the Mendelian proportions, it would follow that the proportion of homozygous mice in the sub-class would be approximately $\frac{64}{65}$ and of heterozygous ones approximately $\frac{1}{65}$.

But an objective chance interpretation is not possible. We could, however implausibly, start with the assumption that the genetic mechanisms are indeterministic, and we could get as far as saying that the set-up described as GH_1 (regarded as a mice-mating bureau followed by something that sifts out seven-membered families) has an objective chance of producing all-black teams. But the conclusions of the form Prob $[H(gf)] = x$ resist this sort of interpretation: no set-up describable as GF can have an objective chance of *producing* homozygousness or heterozygousness in the black parent, unless we are prepared to postulate a backward objective chance on the analogy of backward causation.

Given that the frequency interpretations, however, are legitimate, how can we use them? Clearly we want to say that if we take a mouse that satisfies condition G, and we find that its offspring are all black, the odds are 64 to 1 on its being homozygous, where these odds are a *simple* probability: we want to say that we have pretty good reasons for believing that this mouse is homozygous. To do this, we have to regard our one experimental mouse as a member of either a large finite actual collection of mice that are both G and F or as a member of an indefinitely extended hypothetical sequence of such mice. We then argue that (provided we do not know anything else about this one mouse that bears upon the question whether it is homozygous or heterozygous) it is reasonable to give to the proposition that it is homozygous a degree of belief equal to the (approximate) proportion of homozygousness in the large finite actual collection to which it is known to belong, or equal to the limiting frequency of homozygousness in the hypothetical collective of which it can be considered a member. That is, we have to use the proportional syllogism in one of the usual and already defended ways.

Some Bayesian arguments, then, are not only mathematically correct in accordance with the probability calculus but also appropriate on some defensible interpretations. But disputes arise when we go beyond the sort of case exemplified here.

It was an essential feature of the argument set out above that we began with definite values for Prob $[H_1(g)]$ and Prob $[H_2(g)]$. As Fisher points out, the situation would be very different if we started with no idea of what the proportions of homozygousness and its opposite were in our original class or collective of black mice.[39] Is there any way in which we can still use Bayes's Rule in these circumstances?

First, we can go through the calculations again substituting variables—say n and $1-n$—for the values $\frac{1}{3}$ and $\frac{2}{3}$ used above for Prob $[H_1(g)]$ and Prob $[H_2(g)]$. We then find that:

$$\text{Prob}\,[H_1(gf)] = \frac{128n}{127n+1},$$

$$\text{Prob}\,[H_2(gf)] = \frac{1-n}{127n+1}.$$

The former of these will be the larger as long as $128n > 1-n$, i.e. as long as $n > \frac{1}{129}$. So at least we can say this: as long as the frequency of homozygousness in black-mice-each-of-which-has-been-mated-with-a-brown-mouse-and-produced-just-seven-off-spring is greater than $\frac{1}{129}$, it is more likely that such a mouse, all of whose offspring are black, is homozygous than that it is heterozygous (where this 'more likely' is a simple probability). Can we go further and say that this holds also as long as the *simple probability* that a black-mouse-mated-with-a-brown-mouse-and-producing-just-seven-offspring is homozygous is greater than $\frac{1}{129}$? Failure to distinguish the different probability concepts would have led us to confuse this issue with the pre-

[39] Fisher, op. cit., p. 20.

ceding one, but they are clearly different, and it is not obvious that the same rule will apply. All we can say, I think, is this: as long as it is reasonable to suppose that the frequency of homozygousness in black-mice-etc., is greater than $\frac{1}{129}$, it is more likely that such a mouse, all of whose offspring are black, is homozygous than that it is heterozygous. Of course, we can set different standards of likelihood as we please. We may want to know not just whether the mouse is more likely than not to be homozygous, but whether the (simple) probability of this is, say, better than 0·9, or better than 0·99, and so on. We find, in fact, that as long as it is reasonable to suppose that the frequency of homozygousness in black-mice-etc., is greater than $\frac{9}{137}$, the odds on the all-black-producer being homozygous are better than 9 to 1. Also, as long as it is reasonable to suppose that this frequency is greater than $\frac{99}{227}$, the odds are better than 99 to 1. (As Lucas notes,[40] they are not so in the case considered above, since the frequency $\frac{1}{3}$ is *less* than $\frac{99}{227}$.) This way of speaking, in terms of what it is reasonable to suppose the frequency of the characteristic in question (here homozygousness) to be in the original reference class or collective (here black-mice-etc.) is an elucidation of the usual technical terminology of 'initial' or 'prior' or '*a priori*' probabilities.

This is, perhaps, as much illumination as we can squeeze out of Fisher's mouse. What we next want to know is whether we can interpret Bayes's first and second theorems along the same lines. The first theorem, as Lucas states it, runs:

If a propositional function has come out true on m out of n occasions that are similar and independent of one another, then the most probable value for the probability of the propositional function is around m/n, provided that all the hypotheses assigning a probability-value to the propositional function are each, apart from the information that the propositional function has come out true on m out of n occasions, as probable as any other hypotheses.[41]

In interpreting this, we can take the 'probability of the propositional function', which the rival hypotheses are about, as an objective chance or as a limiting frequency. They are such hypotheses as 'This coin has a chance of about 0·5 of falling heads at each toss', 'This coin has a chance of about 0·7 of

[40] Op. cit., p. 158. [41] Op. cit., p. 151.

falling heads at each toss', and so on. The 'probable' in 'the most probable value' represents a simple probability: it is this hypothesis that it will be most reasonable for someone to accept after he has made the observation, say, that the coin has fallen heads 70 times out of 100. How we should interpret the 'as probable as any other' in the proviso is obscure; I shall discuss this after we have examined the argument.

The different hypotheses as-to-what-the-probability-of-the-function-is correspond to the hypotheses 'this mouse is homozygous', 'this mouse is heterozygous' in the previous example. Producing the observed result true-on-*m*-out-of-*n*-occasions corresponds to producing an all-black family. The proviso that the hypotheses should be initially equiprobable would correspond to a requirement not actually used in the example, that homozygousness and heterozygousness should be equally common among *G*s (where we actually took the ratio to be first 1:2 and later n:1—n).

Let us take a simple illustration, where there are five alternative hypotheses, as follows:

H_1: The probability that a *K* is *L* is between 0 and 0·2
H_2: ,, ,, ,, ,, ,, ,, 0·2 and 0·4
H_3: ,, ,, ,, ,, ,, ,, 0·4 and 0·6
H_4: ,, ,, ,, ,, ,, ,, 0·6 and 0·8
H_5: ,, ,, ,, ,, ,, ,, 0·8 and 1

(To make these exclusive and exhaustive, assume that the lower value is included in each of the five ranges, and that H_5 also covers the limiting value 1.)

Suppose that out of 100 independent trials of *K*, we have found just 70 that are *L* (so that $m/n = 0{\cdot}7$). Then the theorem says that if H_1, H_2, H_3, H_4, and H_5 were initially equiprobable, H_4 is the most probable after the trial results have been observed.

Then the essential structure of the argument is this. Let us call *R* the result that out of the first 100 independent trials of *K*, just 70 should be *L*. Then the probability of getting *R* by way of H_1 is low, as was the probability of getting an all-black family by way of a heterozygous parent; so is the probability of getting *R* by way of H_2, H_3, and H_5. But the probability of getting *R* by way of H_4 is relatively high, as was the probability of getting an all-black family by way of a homozygous parent. Therefore, given that R has been observed to have occurred, it is likely to have come about by way of H_4.

Can we interpret this argument in a way that resembles the legitimate interpretations of the mouse example? We cannot start with a large set of actual situations in one-fifth of which H_1 is true, and so on. We could think of a large set of groups of hypotheses formally similar to H_1, H_2, etc., so that in one-fifth of the groups the H_1 analogue was true, in one-fifth the H_2 analogue, and so on. Since the H_1 analogues would rarely give rise to R analogues, and likewise the H_2, H_3, and H_5 analogues, whereas H_4 analogues would give rise to R analogues more often, we could end by collecting all the R analogues together and noting that most of them arose from H_4 analogues. Consequently, if we know that in our actual situation R itself (which can be regarded as a member of a class of R analogues) holds, and we know nothing else that is relevant to the truth of H_1, H_2, H_3, H_4, or H_5 except for the formal features which they share with all their analogues, then a standard proportional syllogism will yield a simple probability for the truth of H_4 higher than that for any of its rivals. This interpretation seems possible, but it is very strained and it is not clear how we are going to find the initial large set of groups of hypotheses analogous to the *H*s.

It is more tempting to go over to the range interpretation, and to think not of analogues of the *H*s but of the *H*s themselves in a realm of possibilities. Each possible situation is characterized in these two ways:

(*a*) H_i is true (for $i = 1$ or 2 or 3 or 4 or 5).

(*b*) Out of the first 100 *K*s that are observed, exactly x are L (for $x = 1$ or 2 or . . . or 99 or 100). Then to say that the *H*s are equiprobable is to say that of the whole realm of possibilities one-fifth make H_1 true, and so on. Since the probability of R on H_1 is low, very few of the possibilities for which H_1 is true also have $x = 70$, and the same holds for H_2, H_3, and H_5; since the probability of R on H_4 is relatively much higher, a much higher proportion of the possibilities for which H_4 is true also have $x = 70$. So when we consider the sub-class of all and only those possibilities for which $x = 70$, we shall find that more of these have H_4 true than any of its rivals. Since we know that the actual situation is one for which $x = 70$, and since we know nothing else that is relevant to the truth of one or other of the *H*s, a proportional syllogism applied to this sub-class of possibilities yields a greater simple probability for H_4 than for any of H_1, H_2, H_3, and H_5.

However, talk that reifies possibilities in this way cannot be taken ultimately at its face value. It requires further analysis and explanation in terms of what people do. Possibilities do not exist on their own—let alone have measures on their own—but only as the contents of people's considerings. People consider possibilities, give them different degrees of belief, and (perhaps) may be justified in doing so. The range representation has no independent authority by which it can establish that certain ways of reasoning are sound; rather, we need to show independently that the ways of reasoning it would authorize are sound in order to defend it as a helpful and not a misleading picture. Just as, in discussing conditionals, we saw that a possible worlds analysis would call for a further analysis in terms of suppositions and the way people handle them, so the metrical treatment of possibilities which we have just used to illustrate Bayes's theorem requires a further explanation and defence in terms of how it is reasonable to handle qualified beliefs.[42]

When I was discussing detachment, I said that the essential procedure which the range representation encourages is that of holding *relative* degrees of belief constant when we exclude possibilities that we initially left open, and I suggested that consistency would require this. Much the same is what we need here. We start by apportioning belief equally between the *H*s. We then give to the possibility that H_1 and R will both hold just that fraction of the belief given to H_1 that corresponds to the probability of R on H_1 (that is, to the degree of belief that we should have given to R if we had known that H_1 held, and nothing else relevant). We do the same for the other joint possibilities, H_2 and R, H_3 and R, and so on. Then comes the crucial step. When we find that R does hold, we as it were throw away all the bits of belief given to all the possible situations that would not make R true, but keep the same relative degrees of belief that it was reasonable to give to the various possible situations which would make R true. Of course, this is a rather fanciful description; it is at best an idealization of

[42] Cf. ch. 3, § 8, above.

anything that we could actually do with beliefs on their own without a probability terminology and calculus. But that does not matter; what matters is that this is something which it would be reasonable to do, something which consistency would require, and which we diverge from at the cost of being not quite consistent. Reducing the issue to its bare essentials, it amounts to this: suppose we regarded only four situations as being initially possible, H_1 and R, H_1 and not-R, H_4 and R, H_4 and not-R, and suppose that we had already decided that it was reasonable to believe more strongly that the third of these would occur than that the first would. Then if we learn that R has occurred, is it reasonable to believe that H_4 holds rather than H_1? Yes, because learning that R holds does not alter the relative degree of belief which it is reasonable to give to the first and third possibilities: it only rules out the second and the fourth. We cannot consistently say that it was *previously* reasonable to believe that H_4 and R rather than H_1 and R would hold and not *now* reasonable to believe that H_4 holds rather than H_1. This is the core of the reasoning needed. If this core is sound, we can accept all the rest as the mere addition of a metric and a calculus for which the notion of reasonable degrees of belief leaves room as an idealization. And for this treatment, we need only to interpret the initial equiprobability of the rival hypotheses as a simple probability, derived from no matter what.

I have been making rather heavy weather of the problem of *interpreting* Bayes's Theorem, for I think that this is where the main philosophical difficulty lies when we go from simpler applications of Bayes's Rule, like Fisher's genetic example, to the theorem. Most critics, however, have fastened on a different point.[43] They have questioned whether, when we are discussing alternative hypotheses about, e.g., an objective chance or a limiting frequency, it will be reasonable to assign equal initial probabilities to the hypotheses of any such set (discrete or continuous) as our H_1 to H_5.

The stock reply to this objection is to move to Bayes's Second Theorem, which says (in Lucas's formulation):

If the relative frequency with which a propositional function comes true is m/n, on a number, n, of occasions that are similar

[43] e.g. I. Hacking, *Logic of Statistical Inference*, pp. 198–201; Lucas, op. cit., pp. 154–5.

and independent of one another, then the probability of the hypothesis assigning to the propositional function the probability-value m/n approaches as a limit the maximum value 1, as n increases indefinitely, provided that the probability of that hypothesis, apart from the information that the relative frequency is m/n, does not have a probability-value 0.[44]

In terms of our example, this means that we relax the requirement that the *H*s should be initially equiprobable, and that we narrow down the spread of objective chances or frequencies covered by the favoured hypothesis H_4 (making this not 'between 0·6 and 0·8' but as narrow a band as desired about the value 0·7), and have instead only the requirement that H_4 should have an initial probability greater than 0; we then say that the probability *of* H_4 approaches 1 as a limit as the number of observed trials within which we find the frequency 0·7 increases indefinitely.

More simply, it means that the probability of the hypothesis that-the-objective-chance-or-limiting-frequency-is-about-0·7 is very high if in a large number of independent trials the observed frequency is 0·7, provided that the initial probability of that hypothesis was not 0. However low, but above zero, the initial probability is, there is some large number such that an observed frequency of 0·7 in that number of trials will give the hypothesis a high probability.

Since the proof of this theorem is similar to the proof of the previous one, an analogous interpretation will be in order. The initial probability in the proviso need only be a simple probability. The final probability of the hypothesis is also a simple probability, and the two are linked together by the observation and the requirements of consistency. If one admits that it was reasonable, before making any trials, to give some degree of belief to the hypothesis that this coin had at each toss an objective chance of about 0·7 of falling heads, then one cannot without inconsistency deny, after it has come up heads say 70,000 times in 100,000 tosses, that it is reasonable to give a very high degree of belief to the same hypothesis.

This theorem does, therefore, supply what Bernoulli's Theorem could not, a way of arguing back from an observed frequency to an objective chance or limiting frequency. Its

[44] Op. cit., p. 156.

superiority lies in a point touched upon at the end of my discussion of the Bernoulli Theorem: we can argue from '*P* probabilifies *Q*' to 'not-*Q* probabilifies not-*P*' if we can show also that alternative hypotheses to *P* give a relatively higher chance of not-*Q* than *P* does. The Bayesian pattern of argument takes account of all the different ways of getting the observed result from the holding of each of a set of alternative hypotheses. In terms of the diagram used there, we can say that the Bayesian approach succeeds because it looks at the whole of the right-hand end, not only at what is inside the rectangle for *P*.

The objective chance or limiting frequency hypothesis which may thus be made highly probable will itself establish simple probabilities about future cases: if this coin has, e.g. a 0·7 chance of falling heads at each toss the simple probability that it will fall heads at the *next* toss is 0·7, and that it will fall heads about 700 times in the next 1000 tosses is, by Bernoulli's Theorem, quite high. It looks, then, as if we have here at least the basic skeleton of an inductive argument which is reasonable in a sense determined by the principles to which we have appealed. However, some qualifications need to be stated. First, there is always room for Humean scepticism: the above-mentioned prediction holds only so long as the coin goes on having the same objective chance or limiting frequency as it has been established that it probably now has. This *might* change, just as anything else might: the definitions of both objective chance and limiting frequency make these *present* features of e.g. a coin-plus-tossing-method, not predictive descriptions. Secondly, since the Bayesian approach works essentially with a set of alternative hypotheses, it can be applied only where it is reasonable on other grounds to work with such a set. E.g. if we assume that the coin has *some* objective chance of falling heads at each toss, we can use this approach to decide, perhaps within narrow limits, *what* this chance almost certainly is. Alternatively, if we assume that it has some limiting frequency for heads, we can again decide what this frequency almost certainly is. But that there is *some* objective chance, or *some* limiting frequency, here at all is not established by this method, but has to be assumed at the start.

These qualifications must cast a shadow over any hopes that we might have of using a Bayesian approach either to solve Hume's problem of induction or to reduce the confirmation of hypotheses in general to probabilities handled by the standard probability calculus. But these are larger questions on which I do not at present propose to embark. I have been concerned simply to trace some of the more important ways in which we can, and cannot, move about between probability concepts.

§ 9. *The Status of Probabilities*

We come now to the problem of the ontological status of probabilities, to the question whether each of the main probability concepts has, after all, any correct applications.

For one end of the informal concept, indeed, no such problem can arise. In so far as saying that it is probable that *P* is just saying guardedly that *P*, we have only to recognize that this is something which people do, and that in doing this they are not making any further claims that admit of either defence or rebuttal.

The concept of probability as frequency in a finite class is equally proof against ontological doubts, though for a different reason. If there is a finite class of *A*s, each *A* being a discrete individual, it must have some definite number of members. Some number between this one and zero (inclusively) of these *A*s will be *B*, so long as *B* is any feature precise enough to obey, in the field of *A*s, the laws of non-contradiction and excluded middle. And then the ratio of the second number to the first is the probability, in this sense, of an *A* being *B*.

But all the other probability concepts are exposed to some degree of sceptical doubt. About simple probability an extreme sceptic could say that you either know that *P* or do not know this; if you do not know it, you can have no good reason for believing it; so there can be no degrees of justified belief between 1 and 0. It is difficult to find any satisfactory starting point for an argument for or against a position as extreme as this. One can, indeed, ask the sceptic whether he thinks his own position is analytically true, and, if not, why he thinks he is right to adopt it; and one may hope to ensnare him in some form of self-refutation. On the other hand, this scepticism may be based on the real difficulty of fitting any objective goodness of reasons for believing into a plausible account of what there is in the world: it is analogous to aesthetic or ethical scepticism which is based on the difficulty of finding room in the universe for objective qualities of beauty or goodness or for objective duties or obligations or goodness of reasons for acting. And (still analogously with these scepticisms about the objectivity of values) scepticism about the objectivity of simple probability is strengthened by the relative ease with which we could explain away such a concept as a projection of our natural tendencies to form beliefs in certain circumstances and to handle them in certain ways. The occurrence of much the same tendencies in all human beings could be explained as easily today by reference to evolution and natural selection as it could have been explained two or three centuries ago by divine providence. The tendency to form and handle beliefs in the ways we think reasonable will obviously have had some survival value. Perhaps the main point to be made here is how widespread in its effects any genuine scepticism about simple probability would be. This is not just a separate item like moral obligation or aesthetic beauty, whose objective status can be denied without anything much else being affected. If we deny the objectivity of simple probability, we deny that anyone ever has good but inconclusive reasons for believing anything, and since there are very few beliefs for which we have conclusive reasons, this would commit us to a second-order scepticism about pretty well everything: we might still believe a lot of other things, but we should have to admit that we had no right to do so. Simple probability, then, is either objective (in the sense that sometimes, in certain circumstances, one really does have good but inconclusive reasons for believing something) or else an indispensable element in a conceptual system the abandonment of which would leave us in an intellectually very naked state.

We must, of course, admit that a simple probability is always relative to circumstances and in particular to some mixed state of knowledge and ignorance. But this relativity does not entail subjectivity; the probability is not a mere expression of ignorance; it could still be objectively the case that anyone with just that mixture of knowledge and ignorance was *entitled* to give a certain possibility just such a moderate degree of belief.

Again, an analogy with well-known points about ethics is in order: the doctrine that what is right is relative to circumstances, e.g. to a community's resources and existing pattern of life, is to be sharply distinguished from ethical subjectivism.

Whatever status we give to simple probability will largely determine that given to the supposed logical relation of probabilification or quasi-entailment. This notion had the misfortune to be introduced into philosophy at a time when indefinable properties and intuitions to detect them were in favour, and the doubts that most philosophers now have about, say, Moore's indefinable *good* bear similarly upon an indefinable quasi-entailment. The relation of logical form between premisses and conclusion in, say, a proportional syllogism is visible and unproblematic; but it would be a variant of the naturalistic fallacy to identify the relation of support with this visible formal relationship: the support must be something further that the formal relationship carries with it, much as good-making natural properties carry goodness, and it is to detect this support that we seem to need a special intuition. One difficulty is that entailment itself can be treated more plausibly in other ways than as an intuited non-natural objective relation, and if it goes, quasi-entailment might be expected to go with it. These difficulties can be made to look less acute if we avoid the 'intuitionist' terminology and follow the line suggested in §§ 2, 6, and 7 above, defining probabilification by a conditional statement referring to simple probability. '*P* probabilifies *Q* to degree *x*%' may be taken just to mean 'If anyone knew that *P*, and nothing else which was relevant to *Q*, he would be entitled to give *Q* the degree of belief measured as *x*%'. And then if the objectivity of simple probabilities has been conceded, we could give similarly objective status not exactly to probabilification thus defined (in view of what I have argued about conditionals) but at least to each general proposition which sustains such a conditional, that is, to truths of the form 'Whenever someone knows a proposition of the same form as *P*, and nothing else relevant, he is entitled to give to the corresponding proposition of the same form as *Q* just such a degree of belief'.

On the other hand, we get a neater theory if we start at the other end and make probabilification fundamental, defining

'The simple probability for A of P is $x\%$' as meaning 'A's total state of knowledge and ignorance probabilifies P to degree $x\%$'. It could be argued that despite the non-ordinary terminology and the now unpopular epistemology involved, it is *philosophically* no more of a strain to recognize a relation of probabilification than to recognize 'good reasons for believing in certain circumstances': the latter is only a more everyday concept. Again an ethical parallel is in order: it seems more innocent and more in line with common sense to talk about good reasons for acting than to talk about intuited values or obligations, but *any* form of strict objectivism about ethics is philosophically in the same boat as intuitionism, and the 'good reasons' terminology is only an evasion of the awkward issue.

If we admit simple probabilities and probabilifications, from what sources could we derive them and subject to what constraints? I have already defended the principle of indifference. If P, Q, R, etc., form a set of exclusive alternatives which are all related in a strictly symmetrical way to A's total state of knowledge and ignorance, then A has good reasons for giving an equal degree of belief to each of them. This, I have argued, is acceptable in itself, it has many important applications, and indeed it is essential for deriving a simple probability from anything else (except a probabilification), e.g. from a finite class ratio, a limiting frequency, or an objective chance. But the critics have rightly stressed its limitations: there are many problems to which it is not directly applicable, and where the pretence that it is applicable leads to errors; its use is in place precisely where knowledge runs out, and it is silly to use it prematurely, when more relevant knowledge might have been obtained first. A less well-known principle would be this. Where there is some range of exclusive alternatives (whether discrete or continuous) which are related similarly but not strictly symmetrically to A's total state of knowledge and ignorance, A has good reasons for giving each of them (or each finite segment of a continuous range) some non-zero degree of belief. These alternatives resemble one another sufficiently to make it arbitrary for A to rule one possibility out altogether while leaving others open. We might call this the principle of tolerance. Something like this is needed to provide the starting point for an application of Bayes's Second Theorem. For this purpose, the indeterminacy of the principle ('some non-zero degree') does not matter, and it seems that this principle would be applicable in many cases where the principle of indifference would not. For example, if we can assume that a coin, tossed in the standard way, has some limiting frequency for heads, we are inclined to divide the range of possible frequencies into arithmetically equal intervals, say 0 to 0·2, 0·2 to 0·4, and so on. But we *might* instead divide this range into frequencies between $\frac{1}{1}$ and $\frac{1}{2}$, between $\frac{1}{2}$ and $\frac{1}{3}$, and so on. If we were entitled to apply the principle of indifference to the first set of (five) alternatives, would we not equally be entitled to apply it to the second (infinite) set? But of course we cannot consistently apply it to both at once.[45] But the principle of tolerance can be applied to both sets at once (for we can give somehow diminishing finite degrees of belief to the intervals in the second series); any such assignment followed by suitable observations and a use of Bayes's Second Theorem may establish that it is simply probable in a high degree that the limiting frequency is about such-and-such.

As I have said in § 1, what is called the subjectivist theory of probability is really a minimal doctrine about justified degrees of belief. It does not admit that (at least initially) any definite degree of belief about an isolated possibility is rationally justified, or any specific assignment of degrees of belief between the alternatives of an exclusive set; but it does in effect say that someone is justified in using only *coherent* assignments, ones such that no one can 'make a Dutch book' against him. Can we borrow the positive point made here without its negative associations, and say that it is a constraint upon a set of simple probabilities that they must be coherent in this sense? If we can, then we can take advantage of the work done by the subjectivists in showing that probabilities which are coherent in this way will obey the standard probability calculus.[46]

This point has been disputed, some caution is needed. We cannot say that a man is not rational unless he is willing to place and accept bets at coherent rates: there are plenty of reasons why someone might be unwilling to do this, or again,

[45] Cf. J. M. Keynes, op. cit., ch. 4.

[46] E.g. J. Kemeny, 'Fair Bets andI nductive Probabilities', *Journal of Symbolic Logic*, 20 (1955), 263–73.

might be willing to bet in an 'incoherent' way. Nor can we say simply that a set of rationally justified beliefs must have degrees that correspond to a coherent set of betting quotients. For beliefs in themselves, isolated from any method of measuring their strengths (such as the believer's willingness to bet) do not have degrees that are precise enough to be coherent. What we could say, however, is that to be rationally justified a set of beliefs must be such as to lend itself to measurement by a coherent set of simple probability values. And this is, I think, a plausible constraint to place upon rational beliefs. That it is so may be illustrated by considering what would be glaring violations of this constraint, e.g. the beliefs of a man who thinks that each of the three horses in a race is more likely to win it than not.

I do not suppose, however, that the principles of indifference and of tolerance and this constraint of coherence exhaust the sources from which simple probabilities arise. For one thing, to say that they did would be to settle out of hand the question that I have been deliberately leaving open whether the confirmation of hypotheses is, in general, a different kind of non-deductive support from that to which the standard probability calculus applies.

In § 3 I have argued that there might be an actual indefinitely extended sequence with a limiting frequency which also satisfied the requirement of randomness. There seem to be physical phenomena that exemplify this: e.g. if radium atoms left alone for time *t* were arranged in order in some way so as to form an indefinitely extended sequence, they would exhibit a limiting frequency for decay, and any indefinitely extended sub-sequence would exhibit the same limiting frequency as long as it was not selected in the light of actual decayings. But we also apply the notion of limiting frequency to hypothetical sequences, e.g. to what would happen *if* this penny as it now is were tossed indefinitely, or to what would happen if an indefinite series of radium atoms, just like this one and situated just as this one is, were each left alone for time *t*. Such hypothetical limiting frequencies are of course not strictly objective: the term 'objective probability' is misleadingly extended from finite class ratios and limiting frequencies in actual sequences to these hypothetical cases. But hypothetical limiting frequency statements need not be arbitrary: we have seen how the observation of long finite runs, coupled with Bayesian argument, can support a limiting frequency statement. But it is unnecessary and misleading to take such observations as constituting the *meaning* of a probability statement in Braithwaite's manner: the meaning of a hypothetical limiting frequency statement is built up out of the meaning of the form of statement that ascribes a limiting frequency to an actual indefinite sequence and the meaning conveyed by the conditional construction. These statements get their meaning in one way, their support in another.

Of all the main probability concepts, that of objective chance is the one whose applicability is most open to question. I have argued in § 4 that there is room for a distinctive concept of objective chance only where we suppose there to be some indeterministic process, but that even if we do suppose this we are not committed to recognizing objective chances. A chance is a disposition whose realization is a corresponding frequency. Now, if we took an objective chance to be just what I have called a minimal disposition, the statement that a certain radium atom left alone for time *t* has such and such a chance of disintegrating is synonymous with the statement that if a series of atoms just like this one and in just the same situation were left alone for time *t* they would have such and such a limiting frequency of disintegrating, with a random order of disintegrations and non-disintegrations. That is, if objective chance were a minimal disposition, it would be no more than a hypothetical limiting frequency such as we have already admitted. Such a minimal disposition will presumably have a basis: there will be something about the composition and structure of radium atoms which is contingently correlated with their limiting frequency of decay in time *t* while a different composition and structure of, say, radon atoms is contingently correlated with their very different limiting frequency of decay in the same time. But should we recognize an objective chance or chance-distribution which is an intrinsic feature of each trial or experimental set-up (e.g. each radium atom left alone for time *t*) distinct both from this basis and from the minimal disposition, the hypothetical limiting frequency? I have argued that we should not. As I said in § 4, the reasons which I gave for not making such a move with regard to such dispositions as fragility still apply. On the one hand it is unnecessary to postulate any such non-minimal chance disposition; on the other hand, to do so would bring us into conflict with the Humean principle that there are no logical connections between distinct existences. This second point is a little more difficult to expound in this special context, but it still applies. For suppose that there were an objective chance in this non-minimal sense attached, say, to each radium atom. Then this would be, *ex hypothesi*, an intrinsic physical feature of the total set-up, this-radium-atom-left-alone-for-time-t. Now consider the results of a long finite series of such set-ups, the decayings and non-decayings. Since the postulated feature is a dispositional property, a *chance* that this atom should decay in time t whose value is $1-(1/2^{t/T})$, its presence highly probabilifies (by Bernoulli's Theorem) that the frequency of decay in this long finite series is close to $1-(1/2^{t/T})$. Yet these results, with this frequency, are a distinct existence from the supposed intrinsic physical feature of each set-up. And the probabilification would be a logical connection, though not an entailment. The point can be made more strongly if we go to an indefinitely extended series of trials. The presence of the supposed objective chance in each trial will now *entail* that the limiting frequency in the series of results will be $1-(1/2^{t/T})$; but the set of trials and the set of results are clearly distinct existences. This anti-Humean aspect can be removed in only one of two ways. We can identify the chance with the minimal disposition, i.e. with what is asserted by a hypothetical limiting frequency statement, so that it ceases to be a distinct existence from the results. Or we can identify the chance with what I was calling its basis (with whatever relevantly distinguishes, e.g., radium atoms from those of other elements that either have different half-lives or are not radioactive at all), so that the physical reality thus introduced is only contingently related to the results (by a quasi-causal law); it is only the dispositional style of description by which it is introduced that is logically connected to these results—it is introduced as *whatever is responsible* for this limiting frequency of decay. But either of these reforms would do away with the chance as an intrinsic feature of a trial or set-up distinct from both the frequency and the basis.

How far does this account satisfy Popper's purpose in introducing the propensity interpretation of probability? He wanted a kind of probability that could be ascribed to an individual trial or experiment, not merely to a sequence, but also one that was physically real, not a matter merely of subjective belief or even of good reasons for belief.[47] What I am offering in place of his propensity is a hypothetical limiting frequency, the fact that it is in order to use such a statement as 'If there were an indefinitely extended series of atoms like this one . . .'. Being a conditional, this does not correspond directly to a physical reality, but the propriety of using it is linked with two physical realities, the frequency of decay in exactly similar atoms and the structure of this atom which is connected with that frequency by a contingent quasi-causal law. This probability is not just a matter of belief or even of justified degree of belief. Also, since this conditional statement is about *this* atom, this account partly meets Popper's demand for individuation. We can say that this atom-set-up is such that . . ., but this 'such that' stands in for a description (which may or may not be known to us) which is only contingently related to the frequency result. The demand to which we cannot accede is that this 'such that' should introduce a feature which is logically linked with that result. In particular, if we apply this whole account to dice-throwing (assuming this, for the moment, to be an indeterministic process) we can meet Popper's point that we want to say something different about a throw of an unloaded, regular die inserted into a sequence of throws of a loaded die in which the limiting frequency of sixes is $\frac{1}{4}$. *This* throw is such that if there were an indefinitely extended sequence of similar throws the limiting frequency of sixes would be $\frac{1}{6}$. My account, then, does not quite meet Popper's demands, but it goes as far towards meeting them as other philosophical considerations allow. Of course, if we deal with objective chance itself in this way, there will be every reason for dealing similarly with a more elaborate propensity account such as Mellor's. We can accept propensities as minimal dispositions, and we can presume that they will have bases, but

[47] 'The Propensity Interpretation of the Calculus of Probability', as in n. 22, p. 184, above.

we need not and should not recognize propensities as non-minimal dispositions in the rationalist sense.

What I have been suggesting about the ontological status of probabilities can be summed up by answering the question, 'What (in the way of applications of probability concepts) would an omniscient being see as he looked around the universe?'

He would see finite class ratios. If the physicists are right, he would see limiting frequencies, with randomness, in actual indefinitely extended sequences of repetitions of various sorts of indeterministic process. Though he could not see counterfactual conditional statements being made true by corresponding states of affairs, he might quite properly use hypothetical limiting frequency statements about, e.g., individual atoms, though no doubt such toying with suppositions would be, for an omniscient being, a pointless exercise. He would not see objective chances in any stronger sense than this. He would see people holding and expressing beliefs with varying degrees of confidence, though these degrees would not admit of direct measurement or accurate quantitative comparison. In some cases he would see people prepared to lay bets, and the rates at which they were prepared to do this would provide an improved quantitative comparison between degrees of confidence.

If the sceptic about good but inconclusive reasons is wrong, our omniscient being could see that various propositions were simply probable in certain degrees relatively to specific mixed states of knowledge and ignorance, although, being omniscient, he could also see about each of those propositions either that it was true or that it was false. He could also see that certain formal relationships between premisses and conclusions carried with them various degrees of support or probabilifications. Of simple probability and probabilification he would presumably see one as fundamental (but I don't know which) and the other as reducible to it.

This survey shows that even if every proposition is either true or false, it by no means follows that there are no probabilities with values between 1 and 0, and hence (once again) how misleading it is to think of probability as an alternative to truth and falsehood, and of probability values as lying on a scale whose extremes stand for truth and for falsehood.

On the other hand, if the sceptic about good but inconclusive reasons is right, our omniscient being would see no simple probabilities or probabilifications. But he would still see human beings in general coping with their mixed states of knowledge and ignorance in a fairly systematic way, a way of which probability theory and some of its applications and what I have described as valid transitions between probability concepts are partly a description and partly an idealization. He would see that this method of handling mixtures of knowledge and ignorance is on the whole advantageous to those who use it, and he would recognize the human tendency to use it as the outcome of an evolutionary process of natural selection—unless, of course, he himself had had a more direct hand in its formation.

6

LOGICAL PARADOXES[1]

§ 1. *What is the Problem?*

THERE is a group of paradoxes or antinomies, some well known, others less well known, which includes the Epimenides and other forms of the liar, heterologicality, Russell's class paradox, Richard's paradox, and so on. They have been called the paradoxes of self-reference, but that begs more than one question. We should find, I think, fairly widespread support for each of the following views about them, though fortunately, since not all these views are compatible with one another, they are not all held by the same people.

(1) These paradoxes are unimportant and are easily dismissed, since they arise from trivial misuses of language, such as ambiguity or lack of meaning.[2]

(2) While they may be important for set theory and the foundations of mathematics, and perhaps for the philosophy of language, they are of no interest to the general philosopher.

(3) These paradoxes can be solved or removed by valid proofs in formal logic.[3]

(4) Some of these paradoxes compel us to revise our naïve concept of a set or class.[4]

(5) Some of these paradoxes compel us to distinguish sharply between an object language, a metalanguage, a meta-metalanguage,

[1] The main paradoxes discussed here are collected in the Appendix; variants of them are also quoted in the course of this chapter.

[2] This view, and the following one, are met more often in conversation than in the literature. There is at least a hint of the first in P. F. Strawson, 'Truth', *Philosophy and Analysis*, ed. Margaret Macdonald (reprinted from *Analysis*, 9 (1949)), 260–77, esp. p. 271.

[3] J. F. Thomson, 'On Some Paradoxes', *Analytical Philosophy* (First Series), ed. R. J. Butler, pp. 104–19; also some hints in W. V. Quine, *The Ways of Paradox*, pp. 5–20.

[4] W. V. Quine, *Set Theory and its Logic*, p. 5, and *The Ways of Paradox*, pp. 6–18; A. A. Fraenkel and Y. Bar-Hillel, *Foundations of Set Theory*, pp. 7, 13, 19.

and so on. They show that no consistent language can be semantically closed.[5]

(6) These paradoxes show that the habits of thought and speech which we find natural are incoherent and require revision in some ways; the only problem is the technical one of choosing the least inconvenient reforms.[6]

(7) They show that self-reference is logically or linguistically improper.[7]

(8) These paradoxes divide into two radically different groups, semantic or linguistic on the one hand and syntactic or logico-mathematical on the other.[8]

I shall try to show that all these views are mistaken, though some, perhaps, are more mistaken than others.

Let us begin with the second, because this raises the questions of what a paradox is and what it would be to solve it. Typically, a paradox is an apparently sound proof of an unacceptable conclusion; in most, though not all, of ours the conclusion is unacceptable because it is self-contradictory. What we may call *the reasoning within the paradox* has two branches, and the conclusion of one branch contradicts that of the other. In any version of the simple liar, for example, we seem to be able to argue that if the utterance (or whatever else may be used as the paradoxical item) is true, then it is false, so it must be false, but also that if it is false, then it is true, so it must be true, that is, it must be both true and false. But it is also important to consider when and where these antinomies arise. Some of them were introduced into philosophy by the Megarians in the fourth century B.C., some were devised by medieval logicians, some of them live an independent life as children's puzzles; but the special attention given to them over the last seventy years has been due to the way in which they cropped up within the attempts to formalize set theory and the foundations of arithmetic and to describe exact language structures, stating precise rules for determining truth and falsehood. It is easy to see how antinomies, even if their subject-matter appears frivolous,

[5] A. Tarski, 'The Semantic Conception of Truth', e.g. in *Readings in Philosophical Analysis*, ed. Feigl and Sellars, pp. 52–84.

[6] W. V. Quine, *The Ways of Paradox*, p. 18.

[7] A. Ross, 'On Self-Reference and a Puzzle in Constitutional Law', *Mind*, 78 (1969), 1–24.

[8] F. P. Ramsey, *Foundations of Mathematics*, pp. 20–1.

threaten such projects. If the formation rules for a formal system permit the construction of items which the other rules of the system then require to be characterized in incompatible ways, then as long as the system recognizes the validity of the propositional calculus argument-form '*P*, not-*P*, therefore *Q*', any well-formed formula at all becomes provable within the system. Proof within the system ceases to discriminate, and it will be impossible to use any interpretation of the system for any ordinary purpose. Consequently anyone who is constructing a formal system has a vital interest in ensuring that the formation rules and derivation rules that he puts into his system should not, between them, generate antinomies. He wants to *exclude* the paradoxes, to guarantee that none of the known ones will arise within his system, and, if possible, that no as yet unknown ones will break out there either. But the general philosopher or informal logician has an interest of a different sort. The paradoxes also constitute a radical challenge to the rationality of human thinking: they are items about which it is difficult to say anything comprehensive without ourselves falling into contradiction. If we are unwilling to adopt a general scepticism about reason, we must either take up the challenge ourselves or hope that someone else has done so or will do so on our behalf. But this no longer means to show how the paradoxes can be excluded from a formal system, but to show how we can comment on them, wherever they arise, without being committed to contradicting ourselves. The general philosopher wants not merely to keep the antinomies out of this or that intensively cultivated area, but to be able to look them calmly in the face when he encounters them in the wildernesses where they are at home. For the constructor of formal systems, a solution need only be an exclusion device, but for the general philosopher it must be something quite different. It must show that these are only apparent antinomies, perhaps by showing that the issues about which we are tempted into formal contradictions are insubstantial; it must show how our ordinary resources of thought and language allow us to construct paradoxes; it must enable us to understand these contradictions without being ensnared in them. Also, the general philosopher's resources for solving the paradoxes are in some respects more limited. He is not in a position, as the system constructor is, to ban this or to lay down that; he has to comment on what is there already, and he must try to make his comments rationally defensible, not *ad hoc* or arbitrary. Although there may be a wide choice between possible exclusion devices, it seems unlikely that there will be any real choice between philosophical solutions: it is hard enough to find even one. In particular, the simple, general, notion of truth discussed in Chapter 2, which is central in our thinking, is threatened not only by the paradoxes themselves but also by one of the types of solution that has appealed to formal logicians. For these reasons, the general philosopher has an interest in the paradoxes which does not coincide with that of the constructor of formal systems, and he cannot rely on the formal logician to do his work for him.

While all the paradoxes of our group have some resemblances to one another, they also cover a fairly wide range of topics—truth and falsehood, class-membership, definition, and so on—and they are constructed in several different ways. I shall be arguing later (especially in § 5) against view (8) above, that they fall into two distinct groups; but I shall assume throughout that it will be a merit of a proposed solution if it can cope in a fairly unitary way with this full range of paradoxes, and a weakness if it cannot. It is worth noting how readily we can construct variants of the paradoxes—often more or less analogous variants of different paradoxes—for an initially promising solution can often be shown to be worthless by reference to some variant which is immune to it. For example, there are such indirect variants of the simple liar as a card on one side of which is written 'The statement on the other side of this card is false' while on the other side is written 'The statement on the other side of this card is true'. Since there is, strictly speaking, no self-reference here, this variant shows that self-reference, at least in the most literal sense, is not the key to the paradoxes. Again, there are imperative paradoxes like the Sancho Panza, and imperative variants of others. Then there are what I shall call truth-teller variants of all the paradoxes. Corresponding to the simple liar is the simple truth-teller, introduced, for example, by the remark 'What I am now saying is true'. Corresponding to Russell's class paradox is the truth-teller variant, 'Is the class of all and only those classes that are members of themselves a member of itself or not?' Some philosophers will be reluctant to call these paradoxes, since they generate no contradictions. If the truth-teller's statement is true, then it is true; similarly if the class in question is a member of itself, it is a member of itself. But there is still a puzzle here. It is equally arguable that if the truth-teller's statement is false, then it is false, and that if the class of self-membered classes is not a member of itself, then it is not a member of itself. These conclusions do not conflict with the previous ones, but, confronted with both coherent and self-supporting lines of reasoning, we have no way of choosing between them. We cannot decide whether the truth-teller's remark *is* true or false, or whether this class *is* a member of itself or not. In the ordinary paradoxes we have both undecidability and contradiction; in the truth-teller variants undecidability without contradiction. The comparison between them will throw light on the original paradoxes and (as I shall show in §§ 4 and 7) it casts doubt on one otherwise plausible approach to the paradoxes.

Again, there are what I shall call *M*-variants of many paradoxes: the *M*-variant of the simple liar, for example, is introduced by the remark 'What I am now saying is either false or meaningless'. These variants will be used to criticize one kind of solution in § 2, and they will come up again in § 9, where they will contribute to the versions of these paradoxes which are the hardest of all to solve.

Finally, there are the members of what I shall call Prior's family of paradoxes.[9] The ancestor of this family is the oldest of all our paradoxes, the original liar or Epimenides, introduced by the story that Epimenides, the Cretan wise man, said 'All Cretans are liars'. If this means that everything said by any Cretan is false, what becomes of this remark itself? At first sight, there is an easy way out. What Epimenides is taken to have said cannot be true, but it can be simply and non-paradoxically false, and we are forced to move over to the simple liar in order to get something more firmly paradoxical.

[9] A. N. Prior, 'On a Family of Paradoxes', *Notre Dame Journal of Formal Logic*, 2 (1961), 16–32; also 'Some Problems of Self-Reference in John Buridan' (*Proceedings of the British Academy*, 1962), reprinted in *Studies in Philosophy*, ed. J. N. Findlay, pp. 241–59; also *Objects of Thought*, pp. 79–97.

But as we shall see in § 7, the Epimenides has a hidden subtlety of its own, which is exploited in various ways by the later members of this family.

§ 2. *Meaninglessness, Ambiguity, and Vagueness*

It is tempting to say that the sentence used to formulate the simple liar, 'What I am now saying is false', lacks meaning, and is therefore neither true nor false, and similarly that the question 'Is "heterological" heterological?' is meaningless, so that the alleged paradoxes do not arise.

This is tempting because if such charges could be made to stick, it would be a philosophically satisfactory way of dealing with these paradoxes. If something lacks meaning, it cannot be either true or false, let alone both; the reasoning within the paradox is undermined, and the contradiction is thereby removed. Human reason would be vindicated; it is enough that it should be able to see through the pretensions of these meaningless formulations. The builder of formal systems would, no doubt, still need to take precautions against the appearance, within his system, of such meaningless and apparently contradictory items, but whatever revision of his formation rules he introduced as such a precaution would have been given a deeper, extra-systematic, justification. A formal system is, of course, initially uninterpreted and without meaning, but it is intended that it should ultimately be interpreted and applied, and there would be no point in leaving within the system an opportunity for the construction of formulae which would be persistently uninterpretable, whether they caused other troubles or not.

Unfortunately, this charge will not stick. 'What I am now saying is false' is a perfectly constructed grammatical sentence, it is not gibberish, each word and phrase in it is used in its standard way with its standard meaning, and there is not even anything like a category mistake. 'What I am now saying' introduces an item of the right category to be described as true or false, just as would 'What Peter said a moment ago'. Indeed this same sentence 'What I am now saying is false' could be not only grammatically correct and meaningful but even true if it occurred parenthetically, as an aside within some body of speech, and 'What I am now saying' referred to this immediate context.

But, the critic may reply, that makes all the difference. 'What I am now saying is false' is meaningful if its opening phrase refers to anything other than what this sentence is itself being used to assert. But if the supposed statement refers to itself, then it has no meaning.

But this cannot be right. The indirect variant mentioned above would show that we could use this as a general solution only if we were prepared to say that 'The statement on the other side of this card is false' lacks meaning even if something is written on the other side of the card, but it is the wrong thing that is written there. That something like self-reference is the key to these paradoxes is indeed likely, and I shall consider this suggestion further at the end of § 3 and in § 8; but whatever it does, it does not deprive the paradoxical items of meaning. The liar utterance does not lack sense; nor is there even any failure of reference. The reference of the phrase 'What I am now saying' might indeed be uncertain; it might not be clear whether the speaker was referring to this utterance or to his other recent contributions to the conversation. But the paradox is founded not on the ambiguity but on one way of removing it. If this phrase refers to something else in the neighbourhood, there is no paradox; there is a paradox just when this phrase is construed as referring unambiguously to the utterance of which it is a part. And so with all the paradoxes of this group: the reasoning within each paradox requires a precise and unitary way of construing the linguistic components.

Someone might object that we cannot claim at once that 'What I am now saying' has something to refer to and that 'is false' is applied to an item of the right category. He might argue that 'false' is to be predicated not of an utterance but only of a statement or proposition, and that this utterance fails to make a statement. So either 'false' is predicated, with a category mistake, of an utterance, or the referring phrase attempts to refer to a statement or proposition, but there is none.[10] What has been said about truth, and, by implication, about falsity in chapter 2 weakens the force of this objection,

[10] Y. Bar-Hillel, *Aspects of Language*, pp. 253–7 (reprinting a paper first published in *Analysis*, 18 (1957)).

but in any case it can be side-stepped by a reformulation: 'This utterance, standardly construed, says something false.' The only referring phrase now has the utterance to refer to, while 'false' is applied to 'something', which is a variable of the right category, since it is also the object of the verb 'says'. The objector might continue in either of two ways. First, he might demand a namely-rider. But this demand is not in order. Although if it were true that this utterance, standardly construed, said something false, there would indeed be something which could in principle be individually specified which was the (or a) false thing it said, it is not a requirement for the meaningfulness of an existentially quantified statement that it should be filled out, or that it should be fillable out, by a namely-rider.[11] Secondly, the objector could say that the reasoning within the paradox no longer goes through. If it is not the case that this utterance says something false, it does not follow that it says something true: it might fail to say something either true or false. This may in the end be an important comment, but what requires stressing at this stage is that if this utterance does fail to say anything either true or false, it is not for lack of meaning. Some other explanation for this failure would have to be found. In any case, both the objector's possible continuations seem to be blocked by a small further reformulation: 'This utterance, standardly construed, says nothing true.' If *this* utterance failed, for whatever reason, to say anything true, this would seem to ensure that, standardly construed, it did say something true; although equally if it did, it could not. So the reasoning within the paradox goes through; but also, since there is not even a 'something' in this formula, there is no shadow of a pretext for demanding a namely-rider.

I have assumed that meaningfulness is ensured where a suitable sense and reference are provided. A verificationist might point out that the paradox formulae are not empirically verifiable. But if he allows meaning to formulae which are not empirically testable but are analytic he must concede that 'What I am now saying is false' is all too meaningful. It is both analytically true and analytically false. The same sort of

[11] The need for namely-riders is urged in G. Ryle, 'Heterologicality', *Analysis*, 11 (1951), reprinted in *Philosophy and Analysis*, ed. Margaret Macdonald, pp. 45–53, and opposed by P. T. Geach, 'Ryle on Namely-Riders', *Analysis*, 21 (1961), 64–7.

calculations that establish that ordinary analytic truths are true and that ordinary self-contradictions are false will establish that this item has both features. I think there are good reasons for adopting a constructive rather than a verificationist theory of meaning, but even if one did adopt the latter we could not condemn the paradoxical items as meaningless.

The charges of ambiguity and lack of meaning fail even in the case to which we are most inclined to apply them, the simple liar. It is hard to see how they could be brought to bear even on the other 'semantic' paradoxes, let alone 'syntactic' ones like the class paradox. The question 'Is "heterological" heterological?' may be obscure, just because this adjective was coined for the sake of the paradox and is not in ordinary use. But it is merely shorthand for a quite ordinary phrase, 'not truly applicable to itself'. So we can expand the question into 'Is "not truly applicable to itself" not truly applicable to itself?' and then there is no obscurity about the meaning of the question —and, incidentally, there is nothing that can be literally called self-*reference*, though the question is asked about self-*application*. It is even more obvious that there is no lack of meaning in the question asked in Russell's paradox about the paradoxical class; and that class itself, not being a linguistic item, is not of the right category to have or to lack meaning.

M-variants provide another way of countering this charge of lack of meaning. If we replace the simple liar utterance with 'What I am now saying is either false or meaningless', the reasoning within the paradox then runs as follows. If this statement is true, then it is either false or meaningless (since that is what it says); but if it is false, then this satisfies one of the alternatives, and so makes it true, and if it is meaningless, then this satisfies the other alternative, and equally makes it true. Thus if it is true, it isn't, so it isn't; but also if it isn't true, it is, so it is. Even after allowing for the possibility that this utterance may be meaningless, we are still committed to the contradictory conclusion that it is both true and not true. Whereas with the original liar the suggestion that the paradoxical item is meaningless seemed—if only it could have been defended—to yield a way of escape from the contradiction, the suggestion that the corresponding item in the *M*-variant is meaningless only leads us back into the circle from which we tried to break out.[12] I shall come back later (in § 9) to the question how such variants are to be handled. For the present all I want to establish is that the suggestion that paradoxical items are meaningless, even if it were not open to the other objections mentioned, would still not in itself resolve the paradoxes: a new paradox can be built on this proposed solution.

In disposing of the charge of meaninglessness, I have incidentally replied also to the weaker charge of ambiguity or vagueness. Of course any linguistic expression may be ambiguous, and pretty well all have some degree of vagueness or indeterminacy. But the question is not whether these items are ambiguous or vague, but whether their ambiguity or vagueness contributes to the paradoxes, so that removing a bit of vagueness, clearing up an ambiguity, will destroy enough of the argument within the paradox to remove the resulting contradiction. This would be so if the two branches of the argument which lead to opposite sub-conclusions necessarily used different senses of the paradoxical item. But this is just not so. Examination of particular paradoxes shows clearly that both arms of the intended argument within the paradox rely on the same construal of the key item. What can be conceded is that the paradoxical items often have also possible non-paradoxical interpretations. This sort of ambiguity may make the paradoxes more puzzling. We might be readier to recognize the paradoxical items for the very queer constructions that they are if the sets of words used to frame them did not also have quite straightforward uses. But this is a peripheral matter. The paradoxes do not trade on such ambiguity. At most, it puts us off our guard. But once we are on our guard, we still have to decide what to say about these items, and at this stage no pleas about ambiguity are in order.

The charge of ambiguity or vagueness may be linked with a hierarchical theory of language. If it is held that every proper predicate must belong to some one type or level, it will be held as a consequence that a predicate-word is being used ambiguously if it is being made to stand for predicates on more than one level at once. For example, a word like 'long' may be said to have first-level heterologicality, to be heterological$_1$. 'Heterological$_1$' can then be said to be heterological$_2$, and the

[12] J. L. Mackie and J. J. C. Smart, 'A Variant of the Heterological Paradox', *Analysis*, 13 (1953), 61–6, and 14 (1954), 146–9.

argument within the paradox, that if 'heterological' is heterological it does apply truly to *itself* fails. This arm of the argument appears to go through only because we are using 'heterological', at its second occurrence, to do duty for both 'heterological$_1$' and 'heterological$_2$', that is, the argument within the paradox trades on an ambiguity between similar predicates on different levels. Similarly 'true' and 'false' need numerical subscripts, and when they are inserted, the paradoxical items (e.g. in variants of the liar) either cease to be paradoxical (if the subscripts are inserted in one way) or become guilty of type violations (if they are inserted in another) which are equivalent to category mistakes.[13]

That is, the charges of ambiguity and meaninglessness might ride on the back of a doctrine of linguistic hierarchies. But if there is ambiguity or lack of meaning, it is of a sophisticated sort: the misuse of language, if any, is of a subtle and not of a trivial kind. And before these charges can be sustained, the hierarchical theory must first be established. It must be shown not merely that these distinctions of level can be drawn but that they must be drawn, that there is some reason for denying that a word or phrase can be used, univocally, on more than one level. We must turn, then, to the hierarchical theory. The charges of ambiguity and lack of meaning might be consequentially sustained, but they have no independent merit as criticisms of the paradox constructions. We may be able to classify some paradoxes as 'semantic' in that they make use of such terms as 'true', 'define', 'apply to', and so on; but we must not take this classification as meaning that these paradoxes arise from any characteristically semantic difficulty.

Another version of these charges is that some paradoxical item is ill-defined. For example, it may be argued that the diagonally defined number *d* in Richard's paradox is ill-defined in one respect. I shall show (in § 6) that while this may be true, it does not resolve the paradox.

§ 3. *Hierarchical Approaches*

Various forms of the Theory of Types[14] will exclude some or perhaps all of our paradoxes from a system on which they are

[13] Whitehead and Russell, *Principia Mathematica*, I, p. 62.
[14] Ibid., pp. 37–65; Quine, *Set Theory and its Logic*, pp. 241–79.

imposed, but they would provide a philosophical solution only if they had some independent rationale and justification. As *ad hoc* restrictions, imposed just because the paradoxes would arise without them, they would do nothing to solve the paradoxes, though they might be a convenient way of excluding them. In fact, the very simplest theory of types, one which says merely that there are individuals, classes of individuals, classes of classes of individuals, and so on, which can be labelled as items of order 0, 1, 2, and so on, and that an item of order n can have as its members only items of orders less than n, has some intrinsic plausibility, and it will defeat, for example, Russell's class paradox. All classes will be non-self-membered, but there will be no class of all the non-self-membered classes. For any collection of classes there is another class that stands outside that collection, namely the collection itself. This view compels us to give up the naïve assumption that every description, including '. . . is non-self-membered', determines a class, but it does so for good reasons, ones which common sense can accept, even though it did not anticipate them.

This, then, might provide a philosophical solution for Russell's class paradox taken on its own. But it will not in itself solve the 'semantic' paradoxes, and we have already seen that they cannot be coped with separately by characteristically semantic criticisms. Their formal similarity to the class paradox makes it implausible to suggest that they and it should be dealt with in radically different ways. And yet any extension or ramification of the theory of types to deal with them is quite unconvincing. Even if a class of order n can have as members only items of orders less than n, there is no reason why the same description, e.g. 'non-self-membered', should not apply in the same sense to them and to it, and in general there is no prima facie plausibility in the suggestion that 'true', 'define', and so on either can or must have a hierarchy of different senses. Again, if we accept objective properties (as contrasted with verbal descriptions) we can construct a property variant of the class paradox (does the property of not applying to itself apply to itself?); this will not be a *semantic* paradox, since it uses no essentially linguistic terms, and yet the extension of the theory of types that would be needed to cope with it has no intrinsic plausibility: there is, on the face of it, no reason why the property of having more than two members, or of being non-self-membered, should not belong to classes of all sorts and yet still be the same property, and equally no reason why the property of not applying to itself should not belong to properties of all sorts. Similarly, there is no independent reason for introducing a hierarchy of types of truth. Even if we did set up orders of statements (or of sentences), those that are about non-linguistic matters, those that are about statements, those that are about statements about statements, and so on, there is no reason why statements of all orders should not be true in the same sense: things can be as they are stated to be. As we have seen in Chapter 2, this general meaning of 'true' can be elucidated and, if desired, expressed in symbols.

It will not do to reply that these remarks merely reflect old-fashioned prejudices or that we have no solid reason for denying, e.g., the hierarchy of truths. If this approach is to provide a philosophical solution of the paradoxes, we must not merely have no solid reason for denying the hierarchies, we must have a positive independent reason for accepting both the distinctions and the rules that go with them. And except in the case of the original class types we clearly do not have this.

A more radical reply is that summed up in view (6) of those listed at the beginning of this chapter: it is a mistake to look for what I call a philosophical solution. Thus Quine seems to hold that the paradoxes arise from certain features of our language and of our thinking which though natural are yet in a way arbitrary. Our habits of thought can change, and will change under the pressure generated by the paradoxes themselves. The absurdity of the principle of class-existence which produces Russell's paradox will eventually become a commonplace. 'We are driven to seeking optimum consistent combinations of existence assumptions . . . Each proposed scheme is unnatural, because the natural scheme is the unrestricted one that the antinomies discredit. . . .'[15] In effect, what we are here offered as a philosophical solution is simply that we should recognize the arbitrariness and dispensability of the principles of thought and language that give rise to the paradoxes. Once we have done this, all that is left is the technical mathematical or logical problem of choosing the best modification of them, that is, of

[15] Quine, *The Ways of Paradox*, pp. 3–20, esp. pp. 14, 18.

choosing the most convenient device for excluding the paradoxes. Once we see that the naturalness of the unrestricted scheme is contingent upon our actual habits of thought, we can happily adopt whatever chosen restriction achieves the desired freedom from antinomies at the least practical cost.

But this move seems, at the very least, premature. Although there is nothing sacrosanct about our ordinary habits of thinking and speaking, we should at least try to understand *how* they produce real or apparent contradictions. We should then consider whether we can within our ordinary ways of thinking accommodate the paradoxes and reconcile ourselves to them as only apparent contradictions. Only if this fails will we need to reform those ways of thinking themselves; but I shall try to show that it does not fail.

The thesis that we need to distinguish an object language, a metalanguage, a meta-metalanguage, and so on, that no consistent language can be semantically closed, has been developed particularly by Tarski.[16] It would be a piece of pure mythology to pretend to find such a hierarchy already present within a natural language, and Tarski does not make such a claim. Does he hold, then, that any natural language such as English is inconsistent and needs to be reformed?[17] Not quite. His view is rather that a natural language has no 'exactly specified structure', and consequently that the question whether it is consistent or not 'has no exact meaning'. But if a language had an exactly specified structure like that which natural languages seem to have, so that it was both what Tarski calls 'semantically closed' and such that the ordinary rules of logic held within it, it would necessarily be inconsistent. The hierarchical distinctions are introduced not as a description of what is already there, but as a requirement that must be satisfied if inconsistency is to be avoided. But since the supposed proof of this is just the paradoxes themselves, these distinctions do nothing to solve the philosophical problem.

Whether this problem can be solved or not, however, Tarski's thesis, that a language which both is semantically closed and

[16] A. Tarski, 'The Semantic Conception of Truth', *Readings in Philosophical Analysis*, ed. Feigl and Sellars, esp. pp. 59–60.

[17] For a discussion of this point, clarifying Tarski's position and referring to a number of more extreme views, see Y. Bar-Hillel, *Aspects of Language*, pp. 273–85 (reprinting an article first published in *Studium Generale*, 1966).

contains the ordinary logical rules must be inconsistent, calls for some examination in view of its bearing on the problem of truth which was discussed in Chapter 2. A language is semantically closed, in his terminology, if it 'contains, in addition to its expressions, also the names of these expressions, as well as semantic terms such as "true" referring to sentences of this language' and if 'all sentences which determine the adequate usage of this term can be asserted in the language'.[18] But what is it for a *language* to be inconsistent? Prima facie, what can be inconsistent is a statement or set of statements or theory, while a language is only a vehicle, a medium in which things can be said, not a set of statements. However, what is meant is that the formation rules of the language permit the construction of sentences which its other rules—especially the meaning-rules for such terms as 'true'—will require us to call both true and not true.

Tarski's thesis then, amounts to no more than the fact that in a language which is semantically closed in the way in which ordinary English, for example, appears to be, such antinomies as the simple liar can be constructed. But why does this matter? Tarski stresses the commonsense point that an inconsistent theory must contain falsehoods, and is therefore unacceptable. But this comment applies to *theories*, not to a *language* which is inconsistent in the sense explained. Still, it would follow that the rules of an inconsistent language would commit those who always obeyed them and who were ready to answer all questions to saying things not all of which can be true. But it is worth noting that an 'inconsistent' language can be used without embarrassment by anyone who steers clear of certain questions, in much the same way that a car which would fall to pieces at ninety miles an hour can be safely driven at more modest speeds.

But must we accept Tarski's thesis in this sense? I think not. Antinomies like the liar can be blocked not merely by taking away from a natural language elements that it appears to contain but also by adding extra elements, or insisting that they were there, unnoticed by Tarski, already. Thus Prior has argued that a method proposed by Buridan in the fourteenth century and more recently by Peirce would achieve this. If we assume that every statement asserts its own truth (whatever else

[18] Op. cit., p. 59.

it may assert as well) we remove the contradictions by blocking one branch of the reasoning within each paradox, for example, in the simple liar, the argument that if the utterance is false, it is true, and hence that it is true. On the hypothesis that the utterance is false, it follows that one part of what it asserts, namely its own falsehood, is true, but another part of what it asserts, its own truth, is on this hypothesis not true. So we cannot conclude that if it were false it would be true, but instead we must say that if it is false, it is false. The other branch of the reasoning within the paradox, that if it is true, it is false, and hence that it *is* false, still stands, but it stands without opposition: there is no longer a paradox. The simple liar utterance has become contradictory within itself, asserting both its own falsehood and its own truth. We can therefore classify it as simply false, and are relieved of any need to make contradictory comments upon it. There is a world of difference between a merely self-contradictory statement and a paradoxical one. The commentator can self-righteously condemn the former; but the latter tempts the commentator into sinning himself. Thus, Prior says, 'a language *can* contain its own semantics . . . provided that this semantics contains the law that for any sentence *x*, *x* means that *x* is true'.[19]

I want to use this point only destructively, as a disproof of Tarski's thesis. I am unwilling to adopt it as a philosophical solution because its range of application is too narrow. Neither it nor anything analogous to it will, so far as I can see, resolve the paradoxes of Richard and Berry. Also, it does nothing to resolve the truth-teller variants of each paradox. For example, the simple truth-teller utterance already asserts its own truth. Buridan's suggestion would merely reduplicate this, while leaving its consequences unaffected: this item is still true if it's true, and false if it's false, but undecidable between the two. An adequate philosophical solution should deal with all the paradoxes of this group and their truth-teller counterparts.

Another criticism of Tarski's thesis is that it is a misleading

[19] A. N. Prior, 'Some Problems of Self-Reference in John Buridan', *Studies in Philosophy*, ed. Findlay, pp. 251–4, esp. p. 254. In *Formal Logic*, pp. 293–300, Prior expounds Leśniewski's 'ontology and mereology' which provide an analogous solution of the class paradox: every object, including classes, is a member ('element') of itself, so there is no such class as Russell's paradoxical one, and statements about such a class are simply and non-paradoxically false.

suggestion about the source of the trouble. Merely refraining from the use of semantically closed languages is not sufficient to prevent the appearance of antinomies. Suppose that there are two languages, L_1 and L_2, neither semantically closed, but each serving as the metalanguage of the other. And suppose that S_1 is in L_1 and reads 'S_2 is false in L_2', while S_2 is in L_2 and reads 'S_1 is true in L_1'. Then if S_2 is false in L_2, S_1 is true in L_1—because what it says is so—and hence, since this is what S_2 says, S_2 is true in L_2. So if S_2 is false in L_2, it is true in L_2; therefore it *is* true in L_2. But equally if S_2 is true in L_2, 'S_2 is false in L_2' is false in L_1, that is, S_1 is false in L_1, so S_1 is not true in L_1, and therefore S_2 is false in L_2. So if S_2 is true in L_2, it is false in L_2; therefore it *is* false in L_2. Thus we can prove that S_2 is both true and not true in L_2, and similarly that S_1 is both true and not true in L_1: we still have a paradox. This, of course, is merely an indirect variant of the liar expressed in terms of languages. To exclude it, we should need not only the rule that no one language can be semantically closed, but also the rule that no circle of languages can be semantically closed: their relations must be hierarchical and therefore open-ended. But then it is plain that it is not the semantic openness or closedness of a *language* that matters, but the possibility of a semantic circularity.

Should we go back, then, to Russell's (or rather Poincaré's) notion that the basic trouble in all these paradoxes is some kind of vicious circle?[20] I think this general notion is correct, though Russell did not succeed in saying exactly what kind of circularity is at fault. As I have already hinted and as I shall show in § 8 there is nothing essentially wrong with an item's referring to itself, but there would be something logically defective about an item's depending on itself. Nothing can really depend on itself, but the paradox constructions purport to introduce items which would do just this.

I shall try to clarify and develop this thesis throughout the rest of this chapter. But at the moment we are examining hierarchical theories, and I want to stress that the reason why hierarchical devices work, why they do indeed exclude the

[20] Whitehead and Russell, *Principia Mathematica*, I, pp. 37–8. 'The vicious circles in question arise from supposing that a collection of objects may contain members which can only be defined by means of the collection as a whole . . . "Whatever involves *all* of a collection must not be one of the collection".' Cf. Quine, *Set Theory and its Logic*, pp. 241–3.

paradoxes, is that they prevent the construction of circles that would result in (direct or inverse) self-dependence. But they are uneconomical devices: in order to exclude circularities they exclude whole methods of construction, most of whose products would not be circular or objectionable in any way. Russell had to introduce the Axiom of Reducibility in order to bring back all the innocent victims of the Theory of Types. We need unrestricted notions of truth—including the truth of sentences—of being true of, of definition, and we want to be able to treat English, for example, as a single language while we are ourselves using it. There is no reason why sentences should not belong to an object language and its metalanguage at once, as long as they do not contribute to any vicious appearance of self-dependence. In short, many things that would violate type rules or linguistic hierarchies are all right in themselves; in the hierarchical approaches they become victims of guilt by association. It is widely recognized that hierarchical rules are, for this reason, clumsy and inconvenient as exclusion devices; it is not so widely recognized that they are philosophically misleading, suggesting improprieties where there are none.

§ 4. *The Logical Proof Approach*

Another method of dealing with these paradoxes is to rely firmly on what can be established by valid formal proofs. J. F. Thomson has given a particularly forceful example of this method.[21] He first proves a 'small theorem': 'Let S be any set and R any relation defined at least on S. Then no element of S has R to all and only those S-elements which do not have R to themselves.' I shall call this the *barber theorem*, because its most obvious application is to the barber paradox: No collection of men contains a man who shaves all and only those men in the collection who do not shave themselves.

As Thomson says, this theorem is a plain and simple logical truth, and so is the barber application of it. But further applications include 'No collection of classes contains a class having as members all and only those classes in the collection which do not have themselves as members', and 'No collection

[21] J. F. Thomson, 'On Some Paradoxes', *Analytical Philosophy* (First Series), ed. R. J. Butler, pp. 104–19.

of adjectives contains an adjective which is true of all and only those adjectives in the collection which are not true of themselves', and so on. In other words, this plain and simple theorem shows that there is no such class as Russell's paradoxical class, no such adjective as 'heterological' is supposed to be, and so on.

But does this proof solve the paradoxes? Surely not. It disposes of the barber, because we have no reason to suppose that there is, and not much reason to suppose that there might be, such a barber as the story requires. But it does not dispose of Russell's paradox or Grelling's, because we still have on our hands a contradiction between the appropriate interpretation of the barber theorem and the prima facie case for saying that since there clearly are non-self-membered classes, there must be a class that contains them all and only them, or for saying that 'not true of itself' or 'not truly applicable to itself' is a clear and meaningful description; there are precise and known rules for its use, and even if there were not, they could be introduced; this is an English adjective (for the distinction between adjectival phrases and adjectives is irrelevant here) of just the sort that the barber theorem says cannot exist, and the coinage 'heterological' is merely shorthand for it. Thomson himself admits a difference between understanding a paradox and the task of clearing up afterwards, and concedes that it is the second that is complicated.[22] But it is not mere clearing up. The first contradiction (between saying that Russell's class, for example, is a member of itself and saying that it is not) is removed by denying the existence of any such class; but a deeper contradiction, between this denial and the prima facie case for the existence of such a class, remains, and until it is resolved we still have a paradox on our hands. The proposed solution has become half of the argument within a deeper paradox.

Let us look harder at the prima facie case for saying that there is such an adjective as 'heterological' is supposed to be. We commonly suppose that we can define words pretty much as we choose, either altering the sense of existing words in English to suit our (perhaps only temporary) purposes or adding new words to the English vocabulary, or to some technical fragment of it, just as we require them. So why should we not introduce

[22] Op. cit., p. 116.

'heterological' with exactly the sense it was intended to have, so that 'is heterological' is synonymous with 'does not apply truly to itself'? 'Because it has been found to lead to a contradiction' is no adequate answer: it would amount to an *ad hoc* embargo on what, for all we have seen, is a normal and permissible move. But in any case, nothing turns upon the use of the word 'heterological' or any introduced synonym of it. If all such were forbidden, the paradox could still be constructed using only the phrase for which it is supposed to be shorthand, 'not truly applicable to itself'. The sense of this phrase, automatically, as a function of the standard senses of its components, is such that it will apply truly to all and only those adjectives or adjectival phrases which do not apply truly to themselves; and it is *there*, in the English language, already, and no doubt has equivalents in most other languages.

But let us first compare 'heterological' with an imperative paradox like the Sancho Panza. The lord of the manor's instructions to the guard to hang all and only those travellers who give false reports of what they will do, when applied to the awkward traveller who says he is going to be hanged, amount to telling the guard to hang this man if and only if he does not hang him. We have no difficulty in seeing that these instructions are in this case empty—though in other cases they are determinate—because the guard's decision has been made (if he were to attempt to obey) to depend inversely on itself. What constitutes obedience, then, purports to have been made inversely dependent on itself, and the result is that there is, in this case, no such thing as obedience. Yet the instructions are still meaningful: their emptiness, there not being anything that could count as fulfilment of them, is a consequence of their meaning. Similarly, we may be given the task of filling in the table opposite in accordance with the following instructions: 'In the column headed "long", put a tick in any line if and only if the adjective in that line is long; in the column headed "short", put a tick in any line if and only if the adjective in that line is short; and in the column headed "heterological" put a tick in any line if and only if you do not put a tick in that line in the column which has the adjective in that line at its head.'[23] While we have no difficulty in obeying these instructions all the

[23] Based on Thomson, op. cit., pp. 111–12.

way down the columns headed 'long' and 'short' and in the first two lines of the third column, in the third line of the third column they amount to 'Put a tick here if and only if you don't put a tick here'; and this clearly fails as an instruction.

	long	short	heterological
long			
short			
heterological			

It is worth noting that similar comments apply to the truth-teller variants. About another awkward traveller who said merely 'I am not going to be hanged on that gallows' the guard's instructions become 'Hang him if and only if you hang him', which still fails as an instruction; it leaves the guard free to act as his own benevolence or malice may dictate. The appropriate instruction if we added to our table a fourth line and a fourth column for 'autological' would be 'Put a tick in any line if and only if you put a tick in that line in the column which has the adjective in that line at its head', and this, in the fourth line, fourth column space would amount to 'Put a tick here if and only if you put a tick here'.

It is entirely comprehensible that meaningful conditional instructions should thus become empty if their conditionality is so used as to make a decision which attempts to fulfil them depend either inversely or directly on itself. Admittedly these imperative counterparts are only an analogy. 'Heterological' and its longhand equivalents are intended to be descriptions, and what they do or do not apply to should be a matter of fact, not of decision. But there is an analogy here. What they apply to, or not, is a consequence of rules, and the rules for the use of 'not truly applicable to itself' take no grip when this phrase is being considered for application to itself. For the same reason why the ticking instructions become empty in certain cases, the corresponding descriptions fail to describe in the analogous cases. It was the existence of, or the possibility of introducing, rules for its application which was the foundation of the prima facie case for the view that there is, or may be, just such an adjective as 'heterological' is supposed to be. That case is undermined by showing that in this particular situation these rules fail to apply substantially.

A conditional rule governing an action can apply only where whatever the action is made conditional upon is otherwise logically independent of that action itself; the rule has to set up a connection, and it cannot do so if a connection is there already, e.g. if what the action is dependent upon is itself or its own absence. Of course, what the action is made conditional upon may be a result or feature it would have if it were performed; for example, the instruction 'Colour this space red if so doing will make the whole picture symmetrical' is by no means empty; but this is very different from making the action conditional upon its own being performed. In an analogous way, a conditional rule governing a description can apply only where whatever the description is made conditional upon is otherwise logically independent of that description; if what the description is to be derived from is logically tied to the applicability of that description itself, e.g. is identical with it or with its negation, then the description is non-derivable. The rule governing the description is like a tie-rod, but now there are no two points between which it can be fitted. To see that, and why, the rules used in both the imperative and the descriptive paradoxes fail to take a grip we do not need to invoke any hierarchical principles, or any special logical truth like the barber theorem; we need only pay attention to what is involved generally in a rule's applying or being followed.

This point is very like one of those made by Ryle.[24] The question 'Is "long" heterological?' can be unpacked, in view of the meaning rules for 'heterological', to give the more explicit question 'Is "long" not long?' But the question 'Is "heterological" heterological?' resists unpacking. Following the same rules gives us only the translation 'Is "heterological" not heterological?' and further applications of the same rule will only insert additional 'not's, and will never bring us to any more explicit question. But putting it in this way must not be taken as the laying down of a requirement—presumably it would be an *ad hoc* one, to block the paradoxes—that all terms of a certain class should be finitely unpackable or eliminable. Nor is the conclu-

[24] G. Ryle, 'Heterologicality', *Analysis*, 11 (1951), reprinted in *Philosophy and Analysis*, ed. Margaret Macdonald, pp. 45–53.

sion to which this line of criticism leads adequately summed up by saying that the paradoxical items are linguistically ill-formed, let alone that they are meaningless. As we have seen, it is all too easy to construct a new paradox out of that sort of solution. The important point is that the non-unpackability merely reveals and illustrates the fact that no real issue is being raised, that in the case of 'heterological' there is nothing for being heterological to be.

Should we agree with Thomson, then, that 'heterological' is not within its own domain of possible application? Those who use this adjective may intend it to be so. Likewise the standard rules of English would give its longhand equivalents an intended domain of possible application which included all adjectives and adjectival phrases, themselves among them. But, we might almost say, these intentions are providentially frustrated. These words and phrases fail substantially to come within their own domain of possible application. Because of the circularity, when we try to assert or deny any of them of itself we raise no real issue.

By thus undermining the prima facie case for the presumption that there is such an adjective as 'heterological' is intended to be (which includes the assumption that it is a member of the class of adjectives within which it is supposed to make a sharp dichotomy) we remove the deeper paradox which was still there after the barber theorem had been applied, the conflict between the appropriate interpretation of that theorem and the otherwise strong linguistic case for that presumption.

Three further points can be made against the formal proof approach and in favour of the other, approximately Rylean, treatment.

First, the barber theorem does not apply directly to the simple liar. An analogous proof could no doubt be constructed to show that if there was something which was being said by someone who uttered the words, 'What I am now saying is false', with their standard meaning and with the intended self-reference, it would be both true and not true, and hence that nothing is being said in these circumstances. But there would be a crying need for a further explanation of this, for something to undermine the prima facie case for the view that this sentence, being used meaningfully and with an appropriate item to refer to, does say something. Such a further explanation can be given in the Rylean way. There is a standard method of unpacking the content of any statement of the form '*S* is true' or '*S* is false'. Since for *S* to be true is for things to be as they are stated in *S* to be,[25] the method is to find out how things are stated in *S* to be, and then '*S* is true' unpacks into the simple assertion of this, and '*S* is false' into its negation. But when this method is applied to 'What I am now saying is false', we get no further. If we take this as the *S* it says to be false, the standard unpacking yields at the first step only 'What I am now saying is not false'. A second step (taking the opening phrase as still referring to the original statement) yields 'What I am now saying is not not false'. And so on. We never reach anything substantial that is being said. In default of there being an independent how-things-are-stated-to-be, there is nothing for the statement's being false to amount to. The same applies, more simply, to 'What I am now saying is true'. This unpacks always and only into itself, that is, it cannot be unpacked at all. This utterance makes no true statement, and no false one, not because it lacks meaning (for it does not) but because even if it were uttered assertively it would assert nothing. By the standard meaning rules for 'is true', it should say 'Things are as they are herein stated to be'; but there is no way that they are herein stated to be, so no issue can arise. Someone may try to sidestep this treatment of the liar by using the formulation 'What I am now saying is not true' and arguing that by our present account this is not true, and therefore it is true after all. But this is still to treat the question of truth too mechanically. 'What I am now saying is not true' cannot be firmly characterized as not true: it cannot be classified with respect to truth, just because there is nothing for its being true, or not true, to be. An argument that starts from the presumption 'Either it's true or it's not true' should not be allowed to get off the ground. It should be noted that this treatment of the liar paradox is based on the simple or classical notion of truth, the one which was defended in Chapter 2 but which Tarski and his followers believe to be proved incoherent by this and similar paradoxes.

Secondly, there is no truth-teller counterpart of the barber theorem. There may well be a village in which a male barber

[25] Cf. ch. 2.

shaves (somewhat superfluously) all and only those men who do shave themselves. Consequently there can be no formal proof that there cannot be a member of a set S which has R to all and only those members of S which have R to themselves. There is, then, no formal proof that there is no such adjective as 'autological' is intended to be; but then we are left with the puzzle that there may well be the adjective 'autological', but not its negation 'not autological'. Also while it will be a simple matter of fact whether the barber in this village shaves himself or not—he may do either without disturbing the above description of him—it cannot be a simple matter of fact, independently decidable, whether 'autological' applies to itself or not. This ought to be decidable on logical grounds, but it isn't. This and all such truth-teller puzzles are almost as embarrassing as their negative counterparts, and the formal proof approach does nothing to resolve them. But the Rylean approach is equally effective here. It is obvious that with 'autological' there is nothing for its being autological to be: calling 'autological' autological would fail as a description just as saying 'Hang him if and only if you hang him' would fail as an instruction. Thirdly, as I shall show in § 7, reliance on formal proofs of much the same sort as Thomson's can, with the paradoxes of Prior's family, involve us in further absurdities which the Rylean treatment will be needed to resolve.

Thomson's discussion raises one further point. He claims that 'all the solutions of (the heterological) paradox which are usually discussed come to the same thing'.[26] But this is not so. Both a type theory and a metalanguage theory make very extensive claims, and they make them with very little justification, whereas Thomson's own account rests on a logical truth and my Rylean one on general considerations about what is required for a substantial issue. As Thomson says, the hierarchical theories have not merely to stratify predicates (or word-properties); 'it is also necessary for them to claim that this decomposition is exhaustive, i.e. that every predicate belongs on some (one) level'.[27] It would be a *consequence* of this sweeping claim that 'no predicate is true of all heterological predicates of any level', that is, that there is no such predicate as 'heterological' was supposed to be. But the sweeping claim goes far

[26] Thomson, op. cit., p. 113. [27] Op. cit., p. 114.

beyond this consequence, and beyond the corresponding consequences related to the other paradoxes. And in so far as it goes beyond them, it is not supported even by the need to escape the antinomies. As Thomson says, 'the suggestion . . . that we regard "*x* is heterological" as undefined for itself as argument is simpler than any hierarchical account and is to this extent preferable'. On the other hand, the Rylean account explains why this is undefined for itself as argument, whereas the barber theorem by itself does not. Thomson admits that Ryle gives such a reason, but thinks it a weakness that 'this reason counts equally . . . for not counting "autological" as (a philological epithet), (which) leaves it quite unexplained why we should get a contradiction in the one case and not in the other'. On the contrary, it is a merit of the Rylean account that it applies equally to 'autological'. Why we get an apparent contradiction in one case and only undecidability in the other is easily explained by the slight differences in the attempted unpacking procedure. Each step in the vain attempt to unpack the content of '"Heterological" is heterological' (or '. . . is autological') introduces an additional negative from the meaning of 'heterological' where it occurs as the subject, whereas no such additional negatives can be distilled from the meaning of 'autological' as the subject of corresponding sentences.

The various solutions agree on *one* point only: they all deny that there is just such a predicate as 'heterological' was supposed to be. But they differ in the further reasons they give for this: the barber theorem or hierarchical principles or the search for content in the crucial cases.

§ 5. *Lack of Content and Syntactic Paradoxes*

The semantic paradoxes, and their truth-teller counterparts, can be thus solved in a philosophical sense by demonstrating the lack of content of the key items, the fact that various questions and sentences, construed in the intended way, raise no substantial issue. But these are comments appropriate only to linguistic items; one would expect that this method would apply only to the semantic paradoxes and not to 'syntactic' ones like Russell's class paradox, which are believed to involve only (formal) logical and mathematical elements.

Yet there is still a problem about, e.g., the class paradox. The barber theorem seems to prove that no collection of classes can contain a class consisting of all and only the non-self-membered (or, as Thomson calls them, ordinary) classes in that collection, and hence that if we try to make the collection that of all classes whatever, it cannot contain a class of all and only the ordinary classes. But this conclusion comes up against the prima facie case for supposing that there is such a class. We can identify some of its members and we can give a description which is at the same time a criterion for membership of it. Once we know what it is to be an ordinary class, why cannot we in principle 'form the assemblage' of all such classes? Or rather, is it not objectively formed already? This case has still to be undermined, if the barber theorem is not to leave us with a deeper paradox still on our hands.

Quine's view, which is probably shared by many set theoreticians, is that this case is merely a prejudice.[28] The general lesson to be drawn from twentieth-century developments is that sets are tricky things to handle, and that we simply have to abandon the natural or naïve view that there is a class for every property, or for every description, or, as Quine puts it, for every open sentence. Certainly if set theory is developed as an axiomatic system, it will lay down rules that determine what sets there are to be, and there are various ways of doing this which will ensure that Russell's paradoxical class is not a set of the system. But, as I have said, this would be only an exclusion device, not a philosophical solution; the general prima facie case for the existence of the paradoxical class still stands, and cannot be brushed aside as a mere prejudice. Why are classes determined by most properties and most open sentences, but not by a few special ones? How can a property fail to mark off those things that possess it from those that do not? Quine is content to say: 'In view of Russell's paradox we know an open sentence of the object language, namely "—($x \in x$)", that determines no set. In view of Grelling's paradox we know a set which is determined by no sentence of the object language; namely, the set of all sentences of the object language that do not satisfy themselves. If a sentence determined this set, the sentence would be "—(x satisfies x)" or an equivalent; and

[28] Quine, *Philosophy of Logic*, p. 45; *Set Theory and its Logic*, pp. 3–5.

Grelling's paradox shows that no such sentence is admissible in the object language.'[29] But it is surely the acceptance of these shortages on the strength of proofs equivalent to the barber theorem, with no further explanation, that could be called naïve, rather than the opposing 'prejudice'.

But what explanation can be given of the set shortage? If classes are just there, and do not wait to be defined or constructed, the Poincaré–Russell objection to 'impredicative' definition would seem to miss the point. As Quine says, 'we are not to view classes literally as created through being specified . . . as increasing in number with the passage of time. Poincaré proposed no temporal implementation of class theory. The doctrine of classes is rather that they are there from the start. This being so, there is no evident fallacy in impredicative specification.' And Quine concludes that 'the ban urged by Russell and by Poincaré is not to be hailed as the exposure of some hidden but (once exposed) palpable fallacy that underlay the paradoxes. Rather it is one of various proposals for so restricting the law of comprehension (which is involved in our naïve notion of a class)

$$(\exists y)(x)(x \in y. \equiv Fx)$$

as to thin the universe of classes down to the point of consistency.'[30]

But whatever defects there may have been in the Poincaré–Russell formulation, I think that Quine's conclusion is the reverse of the truth. There is a fundamental fallacy to be exposed, not a need for an *ad hoc* restriction to thin down the universe of classes to consistency.

Contrary to appearances, considerations just like those that affect heterologicality are relevant when we ask whether the class of non-self-membered classes is a member of itself or not. What would it be for it to be, or not to be, a member of itself? For the class of men to be a member of itself would be for it to be a man, which it clearly and simply is not. But for the class of non-self-membered classes to be a member of itself would be for it to be not self-membered, that is, not a member of itself. And that would be for it not to be non-self-membered, that is,

[29] Quine, *Philosophy of Logic*, p. 53.
[30] *Set Theory and its Logic*, pp. 241–3.

a member of itself. And that . . . and so on *ad infinitum*. The question whether this class is a member of itself, and the two apparent answers, resist unpacking just as obstinately as do their counterparts about 'heterological'. The apparently concrete question raises no substantial issue: there can be no hard fact either way.

The fact that the elements used in the construction of this paradox, notably class-membership, are syntactic rather than semantic is indeed a red herring. What matters is the attempt to say something, or to ask a question which would set up a choice between two attempted sayings, about class-membership, and there is room for a linguistic failing here, and therefore for a solution that exposes a linguistic failing. If we can show that there is nothing for Russell's supposed class's being self-membered to be, then we can show that the supposed statements that it is, and that it is not, self-membered are empty, and the contradiction between them is only apparent. Even an apparent contradiction is, no doubt, a nasty thing for a set theoretician to have around, but it is not anything for a philosopher to worry about.

If an alleged class is determined purely extensionally, by having this, that, and the other item as its members, then it cannot fail to be as real and determinate as they are. But if an alleged class is determined intensionally, by the fact that all and only its members have a certain feature or property, then the reality and determinacy of the class depends on the determinacy of the property. The property of being non-self-membered is in general a quite real property, but it is a derivative one, and it is determinate where and only where there is something for it to be derivative from. There is something for the class-of-men's being non-self-membered to be derivative from, the simple fact that it is not a man. But the only thing that the class-of-non-self-membered-classes's being non-self-membered could be derivative from is (at the first step) its own negation or (at the second step) itself, and neither of these can supply a genuine derivation. Much the same applies to its truth-teller counterpart. The class-of-self-membered-classes's being self-membered could be derivative only from itself, that is, it could not be derivative at all. The property would here become indeterminate, and that is why it fails to produce a dichotomy of *all* classes, including ones determined by it and its negation, into those that possess it and those that do not.

It is this local indeterminacy of the key property that undermines the prima facie case for supposing that there must be such classes as the paradoxical one and its counterpart, and therefore leaves us free to accept the appropriate interpretation of the barber theorem. But I would stress that this criticism rests on philosophical points of no great subtlety and that need no deliberate introduction. We have no need to lay down any vicious circle principle before we can make them; but after we have made them we can, if we choose, incorporate what they show into a principle to be deliberately observed in, say, axiomatizing set theory. That principle will then have these philosophical points as its extra-systematic justification. Contrary to what Quine says, we can detect a fallacy at the core of this paradox.

We can agree, then, that there is no determinate class of all and only the non-self-membered classes, or of all and only the self-membered classes, though there are classes of each sort. But the reason is not that the former would violate the barber theorem—which the latter would not do. Nor—what is practically equivalent—is it because the paradoxes themselves show that we must modify our naïve concept of a class. Nor, as we can now see, is it because a class must be of a higher order than any of its members; though this doctrine has some intrinsic plausibility, it is not necessary in order to solve the class paradox and it is not sufficient to solve the similar semantic ones, and no extension of it that would be sufficient would preserve its intrinsic plausibility. Nor is it because some axiomatic set theory has been carefully constructed so that the existence of the paradoxical set cannot be proved within it; for that would leave the paradoxical class untamed outside that theory. The reason is that the derivative features which one tries to use to determine the supposed classes are not derived, or derivable, at certain points.

But since it is *this* that undermines the prima facie case for the existence of the paradoxical class, we have no reason for giving up the *other* premiss on which that case rests, the presumption that every (determinate) property determines a class. If the property had been all right, the class would have been all right too, as our naïve and natural view would require.

It may be objected that I *have* given up the naïve view of classes or sets. That depends on just how naïve it was. I hesitate to speak for anyone else in such matters, but it seems unlikely that anyone who admitted that the description 'green', say, was a bit fuzzy at the edges would be quite so naïve as to suppose that there nevertheless *must* be a fully determinate class of all and only the green things. Of course, there *could* still be such a fully determinate class, if it so happened that no real objects fell within the penumbra of the property, that nothing actually had the hue which was neither covered nor excluded by the description 'green'. Again, there could be a fuzzy-at-the-edges class of green things, and in many fields we are content to work with fuzzy classes—e.g. the working class, the middle class. But in logical and mathematical set theory we want non-fuzzy classes, and it is natural to expect that only determinate properties can be relied on to mark them out. Quine, of course, prudently—or perhaps imprudently—likes to steer clear of properties and attributes, and to deal with open sentences. We have fewer naïve convictions about open sentences than about properties, but I do not see why anyone should have been so naïve as to suppose that every open sentence determines a class. Consider the open sentence (type or token) 'x is beside this'. Would anyone suppose that this determined a class—and one different from that determined by the open sentence 'x is not beside this'—if the reference of the word 'this' were not tied down? Surely the natural assumption is merely that every *determinate* property or *determinate* description carries a determinate class with it, and that assumption has not been impugned.

Admittedly the kinds of indeterminacy that may be found in 'green' and in 'this' are different from that in 'self-membered'. Study of the paradoxes brings to light unexpected sorts of fuzziness. But this does not mean that we have to modify our natural view of the relation between properties and classes (any more than we had to modify our simple concepts of truth or applicability in order to deal with the liar and 'heterological'), but only that we have to apply to new cases the rules already implicit in our use of that relation.

Again it may be objected that I have surreptitiously made use of the sort of type-distinction I am pretending to do without. I accept, for example, being a man as something definite that needs no further unpacking, but I do not similarly accept being false, or being heterological, or being 'ordinary'—i.e. non-self-membered. These I accept as something definite only where they can be unpacked. Higher-order properties occur only where they arise from some first-order properties or states of affairs. It is true that such a distinction is implicit in my treatment. But *this* type-distinction, if it is so called, has an independent rationale. I have argued in Chapter 2 that the ordinary concepts of truth and falsehood, the ones that are used in the construction of the liar and truth-teller paradoxes, do involve a comparison between what is said and how things are; this carries unpackability or eliminability with it. And the same can be shown about class-membership, whether this is determined extensionally or intensionally. These 'type-distinctions' are not introduced *ad hoc*, simply in order to solve the paradoxes. And they are in any case quite different from the setting up of an infinite hierarchy of types with rules that restrict the possibilities of class-membership or, what is worse, of the application of predicates to subjects.

§ 6. *Cantor's Diagonal Argument and the Paradoxes of Richard and Berry*

We may expect—rightly, as it will turn out—that Richard's paradox can be handled in much the same way as the others. But it has a special interest because of the way it mimics Cantor's diagonal argument to prove that the real numbers are not denumerable. Will the solution of the paradox undermine that argument as well?

Cantor's proof is a *reductio ad absurdum*. What it immediately proves is that the set of non-terminating decimals between 0 and 1 is non-denumerable. It can be stated as follows:

Suppose that the set of non-terminating decimals between 0 and 1 is denumerable, i.e. that they can be correlated one-one with the numerals 1, 2, 3, etc. Then they can be set out in a doubly infinite table as indicated:

Numeral	Decimal
1	$0.\ a_{11}\ a_{12}\ a_{13}\ a_{14} \ldots$
2	$0.\ a_{21}\ a_{22}\ a_{23}\ a_{24} \ldots$
3	$0.\ a_{31}\ a_{32}\ a_{33}\ a_{34} \ldots$
.	.
.	.
.	.
.	.
.	.

Here 'a_{11}' simply represents the first figure in the first decimal, 'a_{12}' the second figure in the first decimal, and so on. (So that these decimals can be correlated with the real numbers, we can regard any terminating decimal, such as 0·5, as continuing within an infinite sequence of 0s, and hence as non-terminating.)

Now we define another decimal, d, thus:

$$d = 0.\ b_1\ b_2\ b_3 \ldots$$

where $b_n = 1$ if $a_{nn} = 2$, and $b_n = 2$ if $a_{nn} \neq 2$. That is, the nth figure in d is 1 if the nth figure in the nth decimal is 2, and 2 otherwise.

Then d differs from the first decimal in the first place, from the second decimal in the second place, and in general from the nth decimal in the nth place. It is therefore different from every decimal in the table.

But it is a non-terminating decimal between 0 and 1, and by hypothesis the table includes every non-terminating decimal between 0 and 1.

Since this is a contradiction, the supposition must be rejected, that is the set of non-terminating decimals between 0 and 1 is not denumerable. Hence the set of real numbers also is not denumerable.

Richard's paradox can be stated analogously.

Consider the set of all definitions in words of real numbers. Each definition consists of a finite sequence of letters, and they can therefore be put into groups of the same length, and the definitions of the same length in each group, which must be finite in number, can be put into alphabetical or lexicographical order. Then, putting the groups into increasing order of length, we have a definite ordering of all the definitions, even though they are presumably infinite in number. This set of definitions is therefore denumerable, and we can set up a table correlating the definitions with the numerals, and also correlating the definitions with the real numbers they define, where each real number is expressed in terms of an integral part with a non-terminating decimal part.

Then let *d* be the real number whose integral part is zero and whose *n*th decimal figure is one if the *n*th figure in the *n*th decimal is two, and two otherwise.

Then *d* differs from every real number in the table, that is, from every real number that is defined in words.

But the above exposition from 'Consider . . .' to '. . . two otherwise' is a definition in words of a real number *d*, so that *d* is a real number that is defined in words. (It will be noted that the definition of *d* was stated in words alone, at some cost in clarity, but not in precision.)

This is a paradox because we have reached a contradiction by apparently valid reasoning from premisses that seem to be beyond question. The problem is not merely to resolve the contradiction, but also to discover whether in so doing we undermine Cantor's proof as well.

What constitutes the paradox is the conflict between the Cantor-like argument that the definitions are non-denumerable (because if they could be ordered, so could the numbers they define, but based on any proposed denumeration there is an additional 'diagonal' defined number which escapes and so subverts the denumeration) and the ordering argument that the definitions can be put into length-plus-lexicographical order and so are denumerable. (There is the minor complication that the correspondence between definitions and defined numbers might be many-one, at some points, rather than one-one: more than one verbal definition may define the same number. But this would not matter. Even if the same defined number occurred more than once in the table, *d* would still be different from all those in the table.)

This conflict seems to be easily resolved. Given any ordered set of definitions, even an infinite one, which does not contain the above-stated definition of *d*, we can indeed construct the table and define *d* as above, and it will be an additional number; but this shows only that the original set did not contain all possible definitions of real numbers in words. So we try to include all possible definitions in the list, the above-stated definition of *d* among them. By the ordering argument, this definition must occur somewhere in the table. Suppose that it is the *r*th definition in the list. Then what figure is to occur in the *r*th decimal place in *d*? By the definition, it is to be one if the *r*th figure in the *r*th definition is two, and two otherwise. But it is itself the *r*th figure in the *r*th definition; that is, it is to be one if it is two, and two if it isn't two. This is a plainly contradictory instruction, so the definition of *d* is incomplete. It fails to determine a number, because it fails to determine the figure in the *r*th decimal place.[31] Consequently this is not after all the definition of a real number, and the Cantor-like argument to prove the set of all (possible) definitions non-denumerable fails. So the ordinary argument, that both definitions and defined numbers are denumerable, wins by a walk-over.

But not quite. For once we remove the proposed definition from the list, it successfully defines a new 'diagonal' number that is not yet in the table, so it is a correct definition after all. So it ought to be in the list. Have we not shown, then, not merely that if this is a definition, it isn't (because of the difficulty about the *r*th place) and therefore that it isn't, but also that if this isn't a definition, then it is, and therefore that it is? Have we not merely removed the most obvious antinomy only to uncover a deeper one beneath it?

What is at this deeper level paradoxical in Richard's paradox and evades our superficial solution is exactly mirrored in Berry's, which is much simpler, but lacks the mathematical interest given to Richard's by its resemblance to Cantor's proof. Take Max Black's variant: 'The least integer not named in this book.' Suppose that the largest integer 'named' (that is, mentioned or used or referred to or defined) elsewhere in the book is 256. Then this phrase appears to 'name' 257. But then 257 is 'named' in the book after all. And then the phrase doesn't 'name' 257 (but, presumably, 258). But if it doesn't 'name' 257 it doesn't 'name' 258 either; so 256 is after all the largest integer 'named' in the book, so this phrase does 'name'

[31] Something like this is said by T. J. Richards, 'Self-Referential Paradoxes', *Mind*, 76 (1967), 387–403, but criticized by Richards himself in a final footnote and by F. Jackson in *Mind*, 80 (1971), 284–5.

257 after all. So if it 'names' 257, it doesn't, so it doesn't; but if it doesn't, then it does, so it does.

Both these paradoxes seem to illustrate Russell's vicious circle principle, that if we try to base the definition of something on a totality to which the something would belong, we may get into trouble.[32] But it is not enough simply to *ban* such definitions: what are we to say about them if someone ignores our ban and frames them?

It is worth noting some non-paradoxical possibilities in this neighbourhood.

First, we could read the Berry-type reference as referring only to *other* 'namings' of integers: then it unambiguously 'names' 257. Secondly, we could read it so that it merely generates a sequence of numbers. The first time you read it, the largest number so far 'named' is 256, so this phrase 'names' 257. But now 257 has been 'named', so the second time you read it it 'names' 258. And so on. Thirdly, we could read it as explicitly self-referring, as giving the instruction, 'Consider the least integer which this remark, among the others in this book, does not invite you to consider'. If this is a paradox it is easily resolved; we can without hesitation set this aside as giving no substantial direction.

Similarly, we could read Richard's definition of d first as referring only to *other* definitions in words of numbers: then it unambiguously defines some new diagonal number. Secondly, we could take it as thus defining a number the first time we read it; we then construct a new table incorporating this definition, add to the definition some indication that it is to be read a second time, then read it a second time as defining a new diagonal number based on the new table; we then construct a third table incorporating also this modified definition; and so on, getting a sequence of new defined numbers. Thirdly, we can, as we saw above, read it as explicitly referring to itself as a member of the table in relation to which it attempts to define a diagonal number: there is then only an easy paradox which is disposed of by our 'superficial' solution.

The deeper level of paradox lies in a reading which is different from all of these. The Berry phrase is obstinately paradoxical not when it refers only to other references, nor when it

[32] Whitehead and Russell, *Principia Mathematica* I, pp. 37–8.

refers at each reading only to references other than itself-at-that-reading, nor when it refers explicitly to itself (as well as to other references), but when it refers generally to all references in that book, so that it is potentially within its own domain of reference but is actually so if and only if it 'names' an integer. And likewise Richard's definition.

At this point we simply have to recognize that, and how, we can construct such obstinate paradoxes, using the linguistic resources that we have. To construct them, we need at least implicit quantification as well as the possibility of self-reference. Also, we need to refer to a property, such as that of being a successful definition, which is a derivative feature in the same way that being true, being true of something, being self-membered, and their opposites are. We can then arrange things so that what something's being a successful definition should be derivative from includes, crucially, whether it is a successful definition or not; the feature of being, or not being, a successful definition then becomes non-derivable and therefore indeterminate in this case.

The philosophical solution of this puzzle consists essentially in seeing how such obstinate paradoxes are constructed, for example, how something's status as a definition is made apparently to depend (inversely) upon itself. Once we have understood this, we are no longer committed to endorsing contradictions ourselves. We do not have to say that the Berry phrase both does and does not 'name' 257, or that the Richard formula both is and is not a definition of a real number. What is by now the plain truth is that the status of each formula is indeterminate. In these special circumstances there is nothing for such 'naming' or defining to be. But these indeterminacies are not due to vagueness or to any lack of meaning: they result from following in a precise manner the intended meaning-rules—which are also the standard ones, except that 'name' is used in a very wide sense—for the terms that they use. Both the apparent contradictions and the demonstration that they are only apparent result from the use of ordinary concepts.

This same kind of complication, of deeper paradox, can also be found in, for example, the class paradox. If we said simply that there is no class of all and only the non-self-membered classes, and equally none of all and only the self-membered classes, the non-self-membered classes would be left as a fully determinate group, and they would then have to form a class (from the point of view of general philosophy, even if not from that of this or that set theory). Here too we can argue that if the supposed paradoxical class doesn't exist, then it does. So our final judgement here is not after all an application of the barber theorem; it is not that there just is no such class as Russell's, but rather that whether there is or not is indeterminate. It is not only that the derivative feature of being self-membered or otherwise is here non-derivable, but also that the very existence or non-existence of such a class as Russell's (or its truth-teller counterpart) is a derivative apparent feature which being made (inversely or directly) self-dependent is made non-derivable, and therefore not a real feature at all.

What bearing does our conclusion about Richard's paradox have upon the diagonal argument in Cantor's proof?

If our 'superficial' solution had been correct, we could have argued as follows. There is a radical difference between Cantor's and Richard's arguments. Cantor's *reductio ad absurdum* starts with the sheer *supposition* that all the non-terminating decimals are already there as an ordered set; this supposition leads to the contradictory result that there is *another* such decimal, the diagonal one; so this supposition can be unequivocally rejected. In Richard's argument the supposition that all the possible verbal definitions of real numbers are already there as an ordered set is supported by the ordering argument; this supposition is not refuted, because the 'definition' of *d*, which appeared to refute it, turns out to be incomplete and therefore not a proper definition; this 'definition', and the 'real number' it purports to define, can therefore be excluded, and the conclusion of the ordering argument still stands. There is no similar incompleteness in Cantor's definition of *d*, since it is not (as Richard's would be if its 'definition' were a definition) based on itself; Cantor's *d* is based on the *supposedly* already ordered set of non-terminating decimals. (We cannot argue that there is a similar incompleteness here on the ground that *if* the decimals formed an ordered set, whichever one turned out to be *d* would be there already, say as the *r*th decimal in the table, and therefore that the definition of *d* would include the incoherent instruction to make its *r*th decimal figure something different from what it is; the proof shows that there cannot be a decimal in the table which may turn out to be *d*; whatever is in the table, *d* is different from every one of them. And there is no prima facie case, analogous to the ordering argument, for supposing that the real numbers can be simply ordered: it is easy to prescribe an ordering of the *rational* numbers between 0 and 1; but not of the reals.) Thus we can resolve the apparent contradiction in Richard's paradox, showing that the ordering supposition here does not lead to the conclusion that *d* both is and is not a real number defined in words, and hence that this supposition need not be rejected, whereas we can keep the corresponding contradiction in Cantor's proof to use, by *reductio ad absurdum*, to refute the supposition that the decimals are denumerable.

But since that solution was superficial and inadequate, these comments need some modification. We cannot say firmly that Richard's 'definition' of *d* is not a definition, but only that its definition-or-not status is indeterminate. This forces us to modify the conclusion of the ordering argument. Any given set of verbal expressions can be put into a definite length-plus-lexicographical order, but neither the set that includes the 'definition' of *d*, nor the set that excludes it, can be taken as the set of all possible verbal definitions of real numbers: we cannot say that the latter set is denumerable, since the set itself is indeterminate. But all of this only makes our comments on the Richard argument less straightforward: it does not introduce any new principles which would bear upon what we have said about Cantor's proof.

Philosophers of mathematics may quarrel with Cantor's proof on other grounds, e.g. if they object for some reason to any talk about infinite totalities or to *reductio ad absurdum* proofs. Whether they would be right to do so is another question. All I have been concerned to show is that there is nothing in the admitted resemblances between Cantor's proof and Richard's paradox to throw suspicion on that proof. Neither our superficial solution nor our final one carries with it any undermining of the *reductio ad absurdum* in Cantor's proof.

§ 7. *Prior's Family of Paradoxes*

To get anything like a paradox out of the remark of Epimenides the Cretan that all Cretans are liars, we must take this as meaning that everything said by any Cretan is false. So construed, the remark cannot be true, but it seems that it can be simply and non-paradoxically false. But this is where the hidden subtlety comes in. As Prior, following Church, points out, if this dictum is false, as it apparently must be, there must be some other, true, Cretan statement. But now we seem to have discovered a logically necessary connection between two distinct occurrences: a Cretan's merely *saying* one thing logically guarantees that some Cretan should say something else. This would violate Hume's principle, which in itself is utterly convincing—and on which, incidentally, I have relied in criticism of the rationalist view of dispositions and propensities and the doctrine of objective chance in Chapters 4 and 5—that there cannot be logically necessary connections between distinct events. There would, of course, be nothing puzzling or anti-Humean in a logical connection between a certain saying's *being true* and some other occurrence; or in a logical connection between a saying's *having reference* and some distinct occurrence (e.g. for any remark about the first man to land on Mars to have reference it is necessary that some man should land there sometime); or in a logical connection between one event and the aptness of some description given to another (e.g. the correctness of the description of one event as the second landing on the Moon requires that just one other landing on the Moon should have been made earlier). But it would be very different from all of these, and very surprising, if Epimenides' merely saying that everything said by any Cretan is false requires the making, and the truth, of some other Cretan remark.

This point is made all the more surprising by Prior's development of it. He gives a formal proof of a thesis, containing a variable operator d, which can be read as follows: if d that for every p if d that p then not-p, then for at least two p, d that p. Giving d different interpretations, we can extract from this an indefinite number of logical truths, e.g.:

> If it is said by a Cretan that nothing said by a Cretan is the case, then at least two things are said by Cretans.

If it is feared by a schizophrenic that nothing feared by a schizophrenic is the case, then at least two things are feared by schizophrenics.

And so on.[33]

Prior also draws our attention to many similar situations. E.g. the paradox of the preface turns upon the remark 'There is at least one false statement in this book'. This, it seems, cannot be false; but it can be true only if some other statement in the book is false. So it looks as if by making this modest admission, which it is plausible, for inductive reasons, to make, an author can logically guarantee that he has made a mistake somewhere else in the book.[34]

More surprising are the results of contraposing such theses:

Unless something else is said by a Cretan, it is not said by a Cretan that nothing said by a Cretan is the case.

Unless something false is said somewhere else in that book, it is not said in a book that something false is said somewhere in that book.

A more complicated example is introduced by the story of the Policeman and the Prisoner.[35] The policeman testifies that

[33] 'On a Family of Paradoxes', *Notre Dame Journal of Formal Logic* 2 (1961) 16–32. Prior's proof, in the Polish symbolism with *Up* as the universal quantifier 'for every *p*', '*Ep*' as the existential quantifier 'for some *p*', '*Ipq*' for 'that *p* is the same thing as that *q*', and '*E*(2+)*p*' for the numerical quantifier 'for at least two *p*s', runs as follows:

1. *C (UpCdpNp)C(dUpCdpNp)(NUpCdpNp)*—from *CUpdpdq* by substitution
2. *C(dUpCdpNp)C(UpCdpNp)(NUpCdpNp)*—from 1 and *CCpCqrCqCpr*
3. *C(dUpCdpNp)(NUpCdpNp)*—from 2 and *CCpCqNqCpNq*
4. *C(dUpCdpNp)(EpKdpp)*—from 3 and equivalence of 'not-none' and 'some', i.e. of 'not-all-not' and 'some'
5. *C(dUpCdpNp)K(dUpCdpNp)(NUpCdpNp)*—from 3 and *CCpqCpKpq*
6. *CK(dUpCdpNp)(NUpCdpNp)(EpKdpNp)*—substitution in *CdqEpdp*
7. *C(dUpCdpNp)(EpKdpNp)*—syllogistically from 5 and 6
8. *C(dUpCdpNp)K(EpKdpp)(EpKdpNp)*—from 4, 7 and *CCpqCCprCpKqr*
9. *CIpqCKdppKdqq*—substitution in *CIpqCdpdq*
10. *CIpqCKdppNKdqNq*—from 9 and *CCpCqKrsCpCqNKrNs*
11. *CKKdppKdqNqNIpq*—from 10 and *CCpCqNrCKqrNp*
12. *CEpqKKdppKdqNqEpqKKdppKdqNqNIpq*—from 11, *CCpqCpKpq* and quantification theory
13. *CEpqKKdppKdqNqE*(2+)*pdp*—from 12, definition of 2+
14. *CKEpKdppEqKdqNqEpqKKdppKdqNq*—substitution in *CKEpdpEqgqEpqKdpgq*
15. *CKEpKdppEqKdqNqE*(2+)*pdp*—from 14 and 13, syllogism
16. *CdUpCdpNpE*(2+)*pdp*—from 8 and 15, syllogism

[34] *Objects of Thought*, pp. 84–7.

[35] Quoted by Prior, 'On a Family of Paradoxes', p. 20, from L. J. Cohen in *Journal of Symbolic Logic*, (1957).

nothing the prisoner says is true, while the prisoner testifies that something the policeman says is true. From the mere fact that each of them says these things—not from their being true—it follows logically, as an interpretation of a formally valid proof, that one of them—*either* of them—must say something else. And hence, by contraposition, if neither said anything else they logically could not both say what they are supposed to say, though each could say what he is supposed to say so long as the other did not. Another example, based on a puzzle of Buridan's, is that it simply cannot be the case that two people, *A* and *B*, each say something true, a third, *C*, says something false, and a fourth, *D*, says that exactly as many truths as falsehoods are uttered on this occasion.[36] And in the circumstances described in the story of Mr. *X* in Room 7, it seems to be logically impossible for Mr. *X* even to think what he is inclined to think.[37]

Prior concludes, though reluctantly, that 'we must just accept the fact that thinking, fearing etc., because they are attitudes in which we put ourselves in relation to the real world, must from time to time be oddly blocked by factors in that world, and we must just let Logic teach us where these blockages will be encountered'. Prior also says that '*D*'s saying what is attributed to him is not more blocked, as far as this logic goes, by the sayings of *A*, *B*, and *C* than their sayings are blocked by what *D* is supposed to say; and if you hear all these people together and then ask yourself "Which of them is it who hasn't really said anything?" there is no more reason for answering "*D*" than there is for answering "*A*", "*B*", or "*C*".' Prior admits that he would like some favouritism here, but doesn't see where it could come from.[38] (He rejects, for good reasons, a language hierarchy that would yield such favouritism.)

These conclusions, as Prior himself makes clear, are highly improbable; and yet they are supported by what seems to be impeccable formal logic. Logic seems to bar, in some circumstances, the occurrence of certain sayings, thinkings, and so on which are distinct from those circumstances and which in themselves, apart from those circumstances, are logically possible and can be coherently described. Here, if anywhere, we have a conflict between formally valid reasoning on the one

[36] Op. cit., p. 20. [37] Op. cit., p. 30; for the story, see Appendix.
[38] Op. cit., p. 21.

hand and common sense, backed by a Humean philosophical principle, on the other.

To resolve this puzzle, let us first ask the blunt question, 'Well, what if the event which "Logic" tells us cannot occur did occur none the less?' What if neither Epimenides himself nor any other Cretan ever said anything else, and yet one day Epimenides remarked (though heaven knows why he should) that nothing said by a Cretan is the case. In these peculiar circumstances, his remark in fact relates only to itself. And what would it then be for it to be true? By the rules, it would be for it not to be true. And vice versa. In these circumstances, both the derivative properties, truth and the absence of truth, are non-derivable. In these circumstances, there is nothing for the dictum's being true to be and nothing for its not being true to be. Contingently, in these special circumstances, Epimenides' remark (though still fully equipped with *meaning*) acquires the lack of content, the failure to raise a substantial issue, as well as the formal contradictoriness-by-the-mechanical-rules, which the simple liar utterance always has. It becomes, as we may say, liar-paradoxical.

Since this is what *would* happen if what 'Logic' tells us cannot occur did occur, it looks as if 'Logic' is assuming that such liar-paradoxicality cannot happen. And this is so. The formal proofs on which Prior relies appeal, naturally enough, to the laws of non-contradiction and excluded middle *as applied to such formulae as would, when interpreted, become, e.g., 'No Cretan statements are true'*. It is being implicitly assumed in the formal procedures that each such item must itself be either true or not true, and not both. But, as we have seen, in special circumstances such an item becomes indeterminate, the ordinary logical laws fail to apply to it because it raises no substantial issue. To rely on the ordinary logical laws, then, or on proofs which incorporate them, is in effect to deny that liar-paradoxicality can occur, and hence to infer that the combinations of circumstances which would produce it are logically impossible. But of course it can occur, both in the simple liar case and, with the help of these contingent combinations of circumstances, in these more complex ones.[39]

[39] To see that Prior's formal proofs have this character, consider the following interpretation of the proof set out in symbols in note 33 above, taking '*d*' as 'It is said by

The conclusions that can properly be drawn, then, are only of the forms:

If it is *not-liar-paradoxically* said by a Cretan that nothing said by a Cretan is the case, then at least two things are said by Cretans.

Unless something else is said by a Cretan, it is not *not-liar-paradoxically* said by a Cretan that nothing said by a Cretan is the case.

And so on.

In particular, *A*, *B*, *C*, and *D* can all say what Buridan's puzzle reports them as saying, and the penalty is merely that *D*'s remark (but not *A*'s, or *B*'s, or *C*'s) *then* becomes liar-paradoxical. This yields the favouritism which Prior would have preferred, but which 'Logic' did not provide.

Note that if we change the puzzle so that *B*'s remark is false instead of true, *D*'s utterance becomes truth-teller-paradoxical: contingently, in these circumstances, it works like 'What I am now saying is true'. 'Logic' will not now forbid *D*'s saying what he is supposed to say, and yet it is as contingently empty in these circumstances as in those in which 'Logic' did forbid it.

The story of Miniac, the world's smallest computer, may serve as a further illustration of these points.[40] What is puzzling about this is that we appear to get logically guaranteed answers

a Cretan that . . .' and writing '*C*' for '(things) said by a Cretan' and '*T*' for 'true'.
1. If no *C* are *T*, then if it is *C* that no *C* are *T*, then not (no *C* are *T*). This holds because the fulfilment of the first if-clause will make true the *C* mentioned in the second.
2. If it is *C* that no *C* are *T*, then if no *C* are *T* then not (no *C* are *T*).
3. If it is *C* that no *C* are *T*, then not (no *C* are *T*)—from 2.
4. If it is *C* that no *C* are *T*, then some *C* are *T*—from 3.
5. If it is *C* that no *C* are *T*, then both it is *C* that no *C* are *T* and not (no *C* are *T*)—from 3.
6. If both it is *C* that no *C* are *T* and not (no *C* are *T*) then some *C* are not *T*. This holds because 'not (no *C* are *T*)' would falsify the *C* mentioned at the beginning of the if-clause.
7. If it is *C* that no *C* are *T*, then some *C* are not *T*—from 5 and 6.
The other steps in the proof merely put together the results in 4 and 7; since it cannot be the same *C* that is *T* and that is not *T*, then if it is *C* that no *C* are *T*, there must be at least two *C*s.

This procedure presupposes that it is either definitely the case, or definitely not the case, that no *C* are *T*, and hence that any saying that no *C* are *T* is either true or not. But this presupposition is just what would fail if there were a liar-paradoxical *C*.

[40] See Appendix.

to all yes-no questions by a method which common sense tells us has no reliability at all.

To resolve this puzzle, let us again ask the blunt question, 'What if an answer that seems to be thus logically guaranteed turns out, as a matter of fact, to be false?' Suppose, for example, I consult Miniac on the question 'Is there life on Mars?' and suppose that the answer at the first stage to this question is 'Yes', that the answer at the second stage (to the question 'Will your present answer have the same truth-value as your previous answer?') is 'No', so that Miniac's final answer is that there is no life on Mars, but that when someone gets to Mars he finds living organisms already there.

In this case, the first answer was in fact true. So the second question in effect asked 'Will the answer to this question have the same truth-value as a true statement?' If the second answer had been 'Yes', it would have amounted, contingently, in these circumstances, to saying 'This (second) answer is true', that is, to a truth-teller-paradoxical remark like *D*'s in the revised version of Buridan's puzzle. But since the second answer was 'No', it amounted, contingently, in these circumstances, to saying 'This (second) answer is false', that is, to a liar-paradoxical remark like *D*'s in the original version. That is, what in fact happened is that the second answer was liar-paradoxical, but the reasoning within the paradox obscured this fact.

If the first answer had been false, the second question would in effect have asked 'Will the answer to this question have the same truth-value as a false statement?' And then the answer 'Yes' would have been liar-paradoxical, and the answer 'No' truth-teller-paradoxical.

Now the reasoning within the paradox here is that the second answer is either true or false, that if it is true, the first answer was false, that if it is false, again the first answer was false; that is, the first answer must have been false. This reasoning has selected the possible (but not realized) situation which would have made the second answer truth-teller-paradoxical, and has rejected the possible (and in fact realized) situation which makes the second answer liar-paradoxical. But as common sense assures us, either situation could have occurred: Miniac is not in fact responsive to the presence or absence of life on Mars. And the 'Logic' which guarantees that an answer to a self-referring question should be truth-teller-paradoxical rather than liar-paradoxical is a spurious logic.

The full range of possibilities is set out in the following table:

Fact	1st answer	2nd answer	Miniac's final answer	Truth-value of 1st answer	Truth-value of 2nd answer	Truth-value of Miniac's final answer
There is life on Mars	Yes	Yes	Yes	*T*	truth-teller paradoxical	*T*
	Yes	No	No	*T*	liar-paradoxical	*F*
	No	Yes	No	*F*	liar-paradoxical	*F*
	No	No	Yes	*F*	truth-teller paradoxical	*T*
There is no life on Mars	Yes	Yes	Yes	*F*	liar-paradoxical	*F*
	Yes	No	No	*F*	truth-teller-paradoxical	*T*
	No	Yes	No	*T*	truth-teller-paradoxical	*T*
	No	No	Yes	*T*	liar-paradoxical	*F*

Of course all the situations represented by the eight lines of the table are possible, since either possible fact may be combined with the penny's falling either way at each toss. The vital column is that giving the truth-value of the second answer. This answer is always tantamount to either a truth-teller or a liar utterance. If the second answer is 'Yes' when the first was true, or 'No' when the first was false, the second is a 'truth-teller', but if the second is 'No' when the first was true, or 'Yes' when the first was false, the second is a 'liar'. But since a truth-teller utterance *can* be either true or false so far as the mechanical rules are concerned, and is merely undecidable, the reasoning within the paradox, assuming that the second answer must be either true or false, in effect says that this answer must be a 'truth-teller'. This reasoning refuses to admit that liar-paradoxicality can occur, but tolerates, or rather welcomes, truth-teller-paradoxicality. This requires that the situation should be that in one of the four lines, 1, 4, 6, and 7; the other four lines are excluded. And, as the table shows, in lines 1, 4, 6, and 7 the final answer is always true. But the possibilities represented by lines 2, 3, 5, and 8 are equally real, and in them Miniac's final answers are false. So the 'logically guaranteed' answers can in fact be false, as common sense tells us. But this possibility is obscured by the reasoning that excludes the possibility of the second answer's being liar-paradoxical. The truth, regrettably, is not that Miniac cannot give false final answers but that it cannot give them without making its second answers liar-paradoxical.

Miniac, then, is an additional member of Prior's family. The reasoning within the paradox mirrors Prior's formal proofs, and the recognition that liar-paradoxicality can occur is what we need to resolve the puzzle. Of course, Prior would not deny that the penny could *fall* as in lines 2, 3, 5, and 8; but what he would say is that Miniac would not then be *saying* that there is (5 and 8) or isn't (2 and 3) life on Mars, for the fact that there isn't, or is such life would logically block this saying. But this is the mystery that we have resolved.

There is a close resemblance between the conclusion quoted above from Prior's 1961 article ('we must just let Logic teach us where these blockages will be encountered'), which is echoed in *Objects of Thought*, 'It is one of the uses of logic that it brings these hard truths home to us',[41] and Thomson's use of the barber theorem which I discussed in § 4. In each case the appeal is to formally valid reasoning, but in each case this appeal leaves something very puzzling still to be explained, and we can achieve further elucidation by an informal consideration of what is involved in a description's being applicable to something, or in a class's being a member of itself, or of the presuppositions of the proof-procedures used. In each case, also, the incompleteness of a treatment based on formal proof alone is brought out by comparison with the truth-teller variants. For example, there is no formal proof analogous to Prior's which would even seem to establish this thesis:

If d that for every p if d that p then p, then for at least two p, d that p.

This would mean, for example, that if a schizophrenic fears that all schizophrenic fears are realized, then at least two things are feared by schizophrenics.

And yet, if nothing else is feared by any schizophrenics, or if all *other* schizophrenic fears are realized, this one is in a curious

[41] *Objects of Thought*, p. 88.

position: if it's realized, then it is, but if it isn't realized, then it isn't, but nothing can decide which of the two holds. In this case the occurrence and non-realization of at least one other schizophrenic fear is required if this fear is to escape a kind of partial but now crucial emptiness which would do away with the issue whether it is realized. And yet since what is to be avoided is only emptiness, not apparent contradictoriness, this point can be made only informally, not by a proof within the predicate calculus.

There is some temptation to deny the parallelism between the original paradoxes and their truth-teller counterparts here. As a counterpart to the paradox of the preface Prior discusses[42] a man who thinks nothing but that he is thinking something true; he says that there is a vicious self-dependence here, and he tries to formalize reasoning to the effect that 'if anyone is to think *truly* that he is thinking something true, there must be *something else* true that he is thinking . . . and if anyone is to think *falsely* that he is thinking something true, there must be *something else* false that he is thinking . . . (and hence) since if anyone thinks anything at all he must do so either truly or falsely . . . there must be something else . . . that he is thinking'. Prior is concerned about the difficulty of establishing the general formulae that would sum up this reasoning, but I would make two different points. First, this quotation shows how readily Prior makes the kind of assumption which, as I have argued, is implicit in all his formal proofs: 'If anyone thinks anything at all he must do so either truly or falsely.' Secondly, we might consider the reply that this man would just be wrong: if he thinks that he is thinking something true but does not think anything else at all, then he thinks falsely. Although we can argue mechanically that if this man thinks truly then he thinks truly, and if he thinks falsely then he thinks falsely, it seems more natural to adopt the second alternative: a self-dependent thought can be more easily condemned as false than vindicated as true. Similarly, suppose that we are confronted with a page which contains no other successfully-referring phrase, but does contain the phrase 'the only successfully-referring phrase on this page'. Although we could argue that if this is a successfully-referring phrase, it is—since

[42] Op. cit., pp. 91–3.

it then successfully refers to itself—but if it isn't, it isn't—since it has nothing to refer to—the second alternative looks the more attractive.

I think, however, that we should maintain the parallelism between liar and truth-teller variants here, and say that the quoted phrase is indeterminate in status as to whether it successfully refers or not, that a man can think that he is thinking something true without thinking anything else, and that a man who does so will think neither truly nor falsely, but indeterminately, in a truth-teller-paradoxical way.

It seems clear that the anti-Humean (and anti-commonsense) aspect of all the paradoxes of Prior's family is to be resolved by seeing how the reasoning within these paradoxes presupposes that liar-paradoxicality cannot occur, whereas it simply can. But it may not be quite so clear that when liar-paradoxicality does occur in these roundabout ways we can comment on it as satisfactorily as we can on its more direct appearance in, say, the simple liar.[43] I shall come back to this question after a further look at the varieties of self-reference.

§ 8. *Varieties of Self-Reference*

A distinction is sometimes drawn between genuine self-reference, such as gives rise to the liar paradox, and spurious self-reference, which is quite harmless, such as we find in 'The sounds I am now producing are faint' or 'This is an English sentence'.[44] However, it does not seem that the one kind is any more *genuine*, as self-reference, than the other. Nor can we say that in the harmless cases a statement is made about the sentence, the sounds, etc., used in making it, whereas in the harmful cases the reference is to the statement itself. 'This sentence, standardly construed, would say something false'

[43] Arthur Prior died in 1969. It is only fair to record that I put forward essentially this solution to him both privately and in an article, 'Conditionally-Restricted Operations', *Notre Dame Journal of Formal Logic*, 2 (1961), 236–43, but his reply (cf. p. 31 of his own article) is that this leaves *us* too, the commentators, and not merely such characters as *D*, thinking paradoxically 'in the only too straightforward sense of contradicting ourselves; and the job of being rigorously rational even about irrationality . . . is just not done'. In the story of Mr. *X*, for example, (see Appendix) it seems that we still have to say that it both would and would not be the case that nothing thought in Room 7 at 6 p.m. was the case.

[44] A. Ross, 'On Self-Reference and a Puzzle in Constitutional Law', *Mind*, 78 (1969), 1–24, esp. 12.

refers only to the sentence, and yet it introduces a paradox. The more important distinction is between different kinds of predicate or description; faintness is a non-derivative property, so the sort of self-dependence that is found in the paradoxes cannot arise here. Being an English sentence is derivative from the grammatically correct use of the various words, but it is not and cannot be, or be made to purport to be, derivative from itself or from its own negation. But falsity, being successfully-referring, being a definition of a certain class, being in conformity with a certain instruction, and so on are all derivative properties and it can be so arranged that they appear to be derivative from themselves or their own negations: such are the topics of the harmful and important kinds of self-reference.

In the strictest sense, reference would require a referring phrase, but we can use the terms 'refer' and 'reference' more widely, so that sentences beginning 'Everything said by a Cretan . . .' or 'Something said by a Cretan . . .' may be said to refer to Cretan statements in general. And then if what one of these sentences makes or purports to make is itself a Cretan statement, we may say that it is partially self-referring, whereas something that begins 'This sentence . . .' or 'This statement . . .' and so on may be totally self-referring. In general an explicit self-reference will be total, but it need not be: consider 'This statement and others . . .' 'This sentence and the previous one . . .' and so on. And in general an implicit self-reference will be only partial; but it is only contingently that an implicit self-reference is a self-reference at all—it is contingent that the present speaker is a Cretan, or is in Room 7, etc.—and what might have been only a partial self-reference could contingently be total, e.g. 'Something written on this page . . .' where nothing else happens to be written on the page in question.

Total explicit self-reference of the harmful kind lends itself to an expansion, which we construct by attempting to unpack the content in the standard way. Thus 'What I am now saying is false' expands into:

'It is false that it is false that it is false that . . .'

The expansion is endless, and the fact that it cannot be completed may be used to bring out the lack of content of the original remark.

Partially self-referring expressions lend themselves to similar expansions, but ones which throw off branches that are not endless. In some, though not all, cases these branches are conjunctive components of what is said. For example, if we use 'C' for Cretan statements, 'T' for true, 'S' for 'All C are T' as said by a Cretan, and 'C'' for Cretan statements other than S, we have:

$$S = \left\{\begin{array}{l} \text{All } C' \text{ are } T \\ \qquad\& \\ \text{It is } T \text{ that} \end{array}\right. \diagup \left\{\begin{array}{l} \text{All } C' \text{ are } T \\ \qquad\& \\ \text{It is } T \text{ that} \end{array}\right. \diagup \left\{ \text{etc.} \right.$$

This yields an endless series of terminating statements, 'All C' are T', 'It is T that all C' are T', 'It is T that it is T that all C' are T', and so on, and one non-terminating expression 'It is T that it is T that . . .' (for ever). Here all the terminating items are equivalent, but the non-terminating item is truth-teller-paradoxical and empty.

Despite this unavoidable incompleteness of even a partially self-referring expression, we should not say that it is necessarily vicious or improper. A partially self-referring expression could be used, and could be recognized as valid and authoritative in that all its finite branches were acceptable. These branches need not be tediously repetitive as in the example given. It makes good sense, for example, to say that a person has the right to know the rights he has, including this one; this is indeed a principle well worth asserting, particularly against illiberal regimes. If we call this principle R, the person in question A, and sum up A's other rights as the right to X, we have:

$$R = \left\{\begin{array}{l} A \text{ has the right to } X \\ \qquad\qquad\& \\ A \text{ has the right to know that} \end{array}\right. \diagup \left\{\begin{array}{l} A \text{ has the right to } X \\ \qquad\qquad\& \\ A \text{ has the right to know that} \end{array}\right. \diagup \left\{ \text{etc.} \right.$$

The terminating branches now give A the right to X, the right to know that he has the right to X, the right to know that he has the right to know that he has the right to X, and so on. After the first two or three these are not likely to be of practical importance, but there is no need to question their validity.

Partially self-referring expressions containing negations or the word 'false' applied to universal quantifications, or un-negated existential quantifications, are more awkward to handle in this way because some or all of the branches are dis-joined rather than conjoined. Thus if we now use 'F' for false, 'S' for 'All C are F' as said by Epimenides, and 'C'' for Cretan statements other than this S, we have:

$$S = \left\{\begin{array}{l}\text{All } C' \text{ are } F\\ \qquad\&\\ \text{It is } F \text{ that}\end{array}\right. \diagup \left\{\begin{array}{l}\text{All } C' \text{ are } F\\ \qquad\&\\ \text{It is } F \text{ that}\end{array}\right. \diagup \left\{\ \text{etc.}\right.$$

which, if we use 'Q' for 'All C' are F' has the form:

Q & (It is F that Q OR It is T that (Q & It is F that . . .))
which reduces to Q & (—Q OR (Q & (—Q OR . . .))).

In this case logical calculation reduces the statement simply to the conjunction of Q with a non-terminating item 'It is F that it is F that . . .' (for ever). But other negative but partially self-referring expressions would be less accommodating. 'A does not have the right to know what rights he has', for example, yields an awkward sequence of &s and ORs which cannot be reduced by cancellation.

Can we ascribe truth or falsity (and the like) to such in-complete because partially self-referring expressions?

A Cretan remark that all C (including this one) are T will be contingently and non-paradoxically false if there is some other, false, Cretan statement. But if all other Cretan statements are true, its partial self-reference becomes crucial: what ought to be the decisive element in its analysis is the non-terminating item in the expansion, 'It is T that it is T that . . .' But this cannot be decisive; there is a little bit of emptiness here, and it matters. For this remark to be true would be for every conjunct in its expansion to be so; but there is nothing that it would be for this non-terminating conjunct to be true, so there is nothing that it would be for the remark as a whole to be true either. So this remark is either contingently false or, contingently, truth-teller-paradoxical. But we can say that when it is truth-teller-paradoxical it is true in so far as it is determinate: everything in its expansion that could be true is so. And if someone insisted that our principle *R* about rights could not be, say, morally or legally valid because it too, as we have seen, has an incomplete and therefore empty item in its expansion, we might withdraw to the claim that it is valid in so far as it is determinate.

Similarly Epimenides' remark that all *C* (including this one) are *F* is contingently and non-paradoxically false if there is some other, true, Cretan statement. But if all other Cretan statements are false, again its partial self-reference becomes crucial. We are inclined to say that if it is false, it must be true, and vice versa. But we should rather say that in so far as it asserts its own falsehood, it, like its truth-teller counterpart, contains a little bit of emptiness; the contradiction is only apparent, since it disappears in the expansion, where we have only the sequence 'It is *F* that it is *F* that . . .' which there is little temptation to call either true or false. So Epimenides' remark is either contingently false or, contingently, liar-paradoxical; but having seen how this paradoxicality arises, we need no longer be either surprised or disturbed by it. What, on the face of it, is rather more surprising is that when this remark is liar-paradoxical, it is also true in so far as it is determinate; but again, once we have seen how this comes about, we shall cease to be surprised.

Again, 'Something said in this book is false' is either contingently true (if there is some other statement in the book which is false) or, contingently, liar-paradoxical (if there is not), while 'Something said in this book is true' is either contingently true (if there is some other statement in the book which is true) or, contingently, truth-teller-paradoxical (if there is not). And in each case where the remark is paradoxical we may say that it is false in so far as it is determinate, since the determinate disjuncts in these expansions will be false.

In the paradoxical cases (of either sort) we cannot now say, as we did with the simple liar and truth-teller, and so on, that no statement has been made, that nothing has been said, or that no issue has been raised. Issues have been raised, and, as we have seen, the expansions include determinate conjuncts or disjuncts which are true or false. But in all the paradoxical cases no *further* issue is raised by that aspect of the expression which is represented by the non-terminating item in the expansion over and above what has been dealt with by the realization or non-realization of the terminating items. There is nothing further for the truth or falsehood of, for example, Epimenides' remark to be, given that it is true in so far as it is determinate.

So far I have been assuming that the partial self-reference is conscious and either is or might be made explicit, that Epimenides, for example, could throw in the phrase 'including this one'. If so, our expansions are fair representations of what is said in every sense, including 'what the speaker intends to convey'. But suppose that Epimenides does not realize that his, too, is a Cretan statement—he has, perhaps, forgotten his nationality—and does not intend his dictum to be even partially self-referring. The expansion does not now represent what Epimenides intends to convey. But it still gives the formal conditions which (in the contingent circumstance that the speaker is a Cretan) would have to be satisfied if his remark were to count as true. So we can still say that while there is something that it would be for Epimenides' remark to be false, there is nothing that it would be for it to be (completely) true. The complication is that what Epimenides intends to convey may well be true. He intends to condemn only a class of statements made by persons other than himself. But it is still not surprising if what someone intends to convey is true, while the remark by whose use he tries to convey this, taken literally, is unable to be true.

These considerations make it even clearer than before that self-reference, even when it is harmful, does not result in lack of meaning, and that there can be no question of a general logical or linguistic ban on self-reference, even of the kind called genuine as opposed to spurious. We cannot, from the point of view of general philosophy, get rid of the troublesome constructions. All we can do is to understand how they are, or can in special circumstances become, crucially self-dependent, how their little bits of emptiness may come to the surface, and how *then* there may be no substantial issue about their truth, and how any apparent contradictions which then arise from a mechanical style of calculation are only apparent.

§ 9. *Conclusion—the Most Obstinate Paradoxes*

It is a fascinating feature of this group of puzzles that we seem to be able to take any proposed solution and build out of it a new paradox. Suppose we have decided to characterize some utterances as indeterminate on the ground that they are liar-paradoxical or truth-teller-paradoxical in the ways I have explained. Then what are we to say about each of the following utterances:

P: 'This sentence, standardly construed, is indeterminate.'
Q: 'This sentence, standardly construed, either makes a false statement or is indeterminate.'

For *P*, the reasoning within the paradox runs that (assuming throughout that it is standardly construed) if it were true, it would be false, and if it were indeterminate it would be true, but that it can be simply and non-paradoxically false. Can we conclude happily, then, that it is false? Surely not. This sentence is totally self-referring, and being indeterminate is a derivative property which is here made (inversely) self-dependent in just the sort of way that we have found in general to produce indeterminacy. If we say that this sentence is false, we can be asked in what its falsity consists, and we can only reply, in its not being indeterminate because it is false. It's false because it's false because it's false because . . . Its falsity has become truth-teller-paradoxical. If self-dependence ensures emptiness, *this* is empty. We have here a conflict between the apparently neat answer given by the internal logic and the result of an informal, externally critical, reflection. But the latter must prevail. We cannot let the reasoning within the paradox here force us to call *P* false any more than we can let the reasoning within one of Prior's paradoxes tell us that certain thinkings, fearings, etc., just cannot occur in certain circumstances.

But if *P* is indeterminate, how do we evade the argument that it is therefore true, since that is just what it says? Turning to *Q*, however, we see that whatever we had said about *P*, we should somehow have had to evade such an argument about *Q*. For the reasoning within the paradox here says that if *Q* is true, it is false or indeterminate, but if it is false it is true and also if it is indeterminate it is true. The internal logic here leaves us no resting place. But we surely want to say that *Q* is indeterminate, so we must block the argument from indeterminacy back to truth.

This seems at first sight to be a difficulty for other approaches as well. Consider:

P′: 'This sentence, standardly construed, violates type restrictions.'
Q′: 'This sentence, standardly construed, either makes a false statement or violates type restrictions.'

If *P′* were true (on any level) it would have to violate type restrictions, and if it violates type restrictions, is it not true? Since a type theorist will deny that a sentence can both violate type restrictions and be true, he may be tempted, by this reasoning within the paradox, to say that *P′* can neither be true nor violate type restrictions, but must be false. But *P′* obviously violates type restrictions. Whatever level it belongs to, since it speaks about a sentence on that level (itself) it must belong to a higher level. It is not enough for the type theorist merely to say that he does not allow such formulae to be constructed. Within his system of formal languages they do not occur, but we want to know what comment he will make on these ordinary language constructions which patently have occurred, and whether he can comment on them without contradicting himself.

He can and will claim that he does not himself use any unrestricted concept of truth as, say, things being as they are said to be, but only an appropriate set of recursively defined truth-predicates, one for each language or language-level, and he can then say that while *P′* violates type restrictions there is no truth-predicate which he is on that account committed to applying to it: its type violation will prevent it from satisfying any properly constructed truth-definition. And if we ask whether it none the less satisfies the informal concept of things being as they are said to be, he will say that this unrestricted concept is incoherent, and that there may well be cases where no one can tell whether it applies or not.

For the moment this gets the type theorist off the hook. But can *we* not deal with the difficulty in much the same way? If *P* is, as our informal argument seems to show, indeterminate, then it does not succeed in saying that it is itself indeterminate, though it seems to. The argument back from the hypothesis that *P* is indeterminate to the conclusion that it is true is incoherent: it is convincing only because we do not hold fast to that hypothesis, but take it along with the incompatible assumption that *P* determinately asserts its own indeterminacy. We inconsistently combine a literal, mechanical, interpretation of its content with the critically reflective judgement that it is indeterminate.[45]

In saying this, am I after all conceding point (6) of those listed at the beginning of this chapter, and in particular giving up the simple, unrestricted concept of truth? I think not. I am still saying that truth consists in things being as they are stated to be; the sophistication enters in our handling of the question, 'How, in this utterance, are things stated to be?' And while I temporarily let the type theorist off the hook a moment ago, I think that his claim that the unrestricted concept of truth is incoherent can only be interpreted in an equivalent way. It is the mechanical handling of this concept that is incoherent; if you want a concept that can be handled mechanically you need a restricted one; but any difficulty in the notion of things being as they are stated to be must lie in some problem about how they are stated to be. And what in the Tarskian tradition is called a definition of *truth* is mainly a clarification of what is stated, and perhaps a restriction of what can be stated.

To say, as I have, that an utterance may fail to state what it appears to state, e.g. its own indeterminacy, is reminiscent of Prior's remark (to which I objected) that thinkings, fearings, etc., may be oddly blocked. But I think there is a difference. The failure of *P*, for example, to say that it is itself indeterminate in such a way that its indeterminacy would make it true is not shown simply by 'Logic'—that is, by the need to avoid formal contradictions, and to let anything that is said obey the law of excluded middle, to be either true or not. It is explained by *P*'s own indeterminacy, and this in turn is explained by its total self-reference and the self-dependence that results therefrom.

P and *Q*, however, are relatively easy to deal with just because their self-reference is explicit and total. As we saw in § 6 with Richard's and Berry's paradoxes, the deepest and most obstinate paradoxes arise not from items that refer explicitly to themselves (even if they refer also to other things) but from

[45] The present puzzle is, of course, very similar to what have been called *M*-variants; see § 2 and n. 12 above. The solution suggested here is essentially the same as that offered in the article referred to in n. 12, on p. 246.

items that refer generally to some class in which they happen to be included, or in which they would be included if they achieved their purpose. Let us therefore combine the present complication with the previous one. Let us consider a remark by Epimenides, still forgetful of his own nationality, 'Every sentence uttered by a Cretan, standardly construed, either makes a false statement or is indeterminate.' Or, what comes to the same thing, 'No sentence uttered by a Cretan, standardly construed, makes a true statement.' Let us assume, what might be the case, that no other sentence uttered by a Cretan, standardly construed, makes a true statement. We cannot without contradicting ourselves allow that Epimenides' remark makes a true statement. And yet if it fails *for whatever reason* to make a true statement, we must ourselves say exactly what Epimenides has said; how then can we deny that this is a sentence uttered by a Cretan which, standardly construed, makes a true statement? How can we now avoid contradicting ourselves? This example, which brings together several of my recent themes, seems to be by far the toughest version of the liar.

It is tempting to look for help from a distinction between what Epimenides intends to convey and what his sentence, as uttered by a Cretan, has as its contingent formal truth-conditions. But though what he intends to convey—that no utterances of those other people that he recognizes as Cretans, standardly construed, make true statements—is non-paradoxically true, this is not the whole of what *we* intend to convey. We are saying that no Cretan utterances at all are true; this is what Epimenides' sentence, standardly construed, would assert. So how can this *sentence* fail to be true, as well as what he intends to convey? In any case we can sidestep this distinction by using an example to which it does not apply. Consider a sheet of paper on which are written just two sentences, 'Two and two make five' and 'There is no sentence written on this sheet which, standardly construed, would make a true statement'. Again we cannot say that the second sentence would make a true statement, and yet if we say, for whatever reason, that it would not, then, since the only other sentence on the sheet would make a false statement, we are committed to using or endorsing a sentence which is another token of the same type and with the same reference as this one. How can we avoid contradicting ourselves here? Uncritical reliance on formal logic would result in a proof that two and two make five.

Suppose that we expand 'true' here, replacing 'would make a true statement' with 'would state that things are as they in fact are'. And remember that the things in question include the success or failure of this sentence itself in this respect. I think we can and must say that because of the very tricky kind of self-reference and consequent self-dependence in this case, there just is no *how things are* in the key respect. Consequently, we cannot either endorse or deny a sentence-token of the same type and with the same reference as the second sentence on that sheet. We cannot now drive a wedge between what we say about that sentence and what we allow it to say about itself: standardly construed, it would say just what our own use of the same type sentence would. We must just admit that the issue it appears to raise is indeterminate, and hope that our study of self-reference has explained why this is so. This sentence's indeterminacy with respect to truth is of a kind which prevents our saying even that it is not true, and therefore from arguing, by a further step, that it is true. Awkward as this conclusion is, it has the merit of being analogous to what we found it necessary to say about the 'deeper paradoxes' that we identified in those of Richard and Berry.

As Prior says, anyone who writes about the logical paradoxes has the job of being rigorously rational even about irrationality —that is, about irrational thinking, and about states of affairs that seem in themselves to defy reason. I am not sure that I have succeeded. But the rival approaches which I have examined are clearly inadequate, in one way or another. I have given reasons for rejecting all of the views listed at the beginning of this chapter. We need a philosophical solution for all the paradoxes of this group, not merely an exclusion device, and for this something like what I have tried to formulate will be required.

APPENDIX

A COLLECTION OF LOGICAL PARADOXES

1. *Epimenides*

Epimenides the Cretan said 'All Cretans are liars.'

2. *Simple liar* (attributed to Eubulides of Megara)

A man says that he is lying. Is what he says true or false?

Cicero, *Prior Academics*, II, 96.

3. *Variants of the Simple Liar*

What I am now saying is false.

S is defined as follows: S = 'S' is false.

There is a page on which one and only one sentence is printed in brackets, namely 'The sentence printed in brackets on this page is false.'

'Yields a falsehood when appended to its own quotation' yields a falsehood when appended to its own quotation.

Quine, *The Ways of Paradox*, p. 9.

4. *Imperative Variants*

Disobey this order.

(Indirect) Carry out the next instruction; do not carry out the previous instruction.

5. *Heterological*

Some adjectives have meanings which are predicates of the adjective word itself; thus the word 'short' is short, but the word 'long' is not long. Let us call adjectives whose meanings are predicates of them, like 'short', autological; others heterological. Now is 'heterological' heterological? If it is, its meaning is not a predicate of it; that is, it is not heterological. But if it is not heterological, its meaning is a predicate of it, and therefore it is heterological.

Ramsey, *Foundations of Mathematics*, p. 27.

6. *Berry's Paradox*

The least integer not nameable in fewer than nineteen syllables.

The number of syllables in the English names of finite integers tends to increase as the integers grow larger . . . Hence the names of some integers must consist of at least nineteen syllables, and among these there must be a least. Hence 'the least integer not nameable in fewer than nineteen syllables' must denote a definite integer; in fact, it denotes 111,777. But 'the least integer not nameable in fewer than nineteen syllables' is itself a name consisting of eighteen syllables; hence the least integer not nameable in fewer than nineteen syllables can be named in eighteen syllables.

Whitehead and Russell, *Principia Mathematica*, I, p. 61.

The least integer not named in this book. Max Black, *The Nature of Mathematics*, p. 98.

7. *Richard's Paradox*

Consider all decimals that can be defined by means of a finite number of words; let E be the class of such decimals. Then E has $\aleph_0$ terms; hence its numbers can be ordered as the 1st, 2nd, 3rd, ... Let N be a number defined as follows: If the nth figure in the nth decimal is p, let the nth figure in N be $p+1$ (or 0, if $p = 9$). Then N is different from all the members of E... Nevertheless we have defined N in a finite number of words, and therefore N ought to be a member of E. Thus N both is and is not a member of E.

Principia Mathematica, I, p. 61.

8. *Sancho Panza* (*or The Gallows*)

A certain manor was divided by a river upon which was a bridge. The lord of the manor had erected a gallows at one end of the bridge and had enacted a law that whoever would cross the bridge must first swear whither he were going and on what business; if he swore truly, he should be allowed to pass freely; but if he swore falsely and did then cross the bridge he should be hanged forthwith upon the gallows. One man ... swore 'I go to be hanged on yonder gallows', and thereupon crossed the bridge.

The vexed question whether the man shall be hanged is brought to Sancho Panza as governor of Barataria.

Cervantes, *Don Quixote*, Part 2, Chapter 51, quoted by A. Church, *Introduction to Mathematical Logic*, p. 105.

9. *Euathlus*

Protagoras agreed to teach Euathlus rhetoric, on condition that Euathlus would pay him a certain sum of money when he won his first court case. But after completing the course, Euathlus did not engage in any lawsuits. Growing impatient, Protagoras sued Euathlus for payment of his fee. He argued: 'If I win this case Euathlus will be bound to pay me, for the court will have so decided; if I lose it, Euathlus will still be bound to pay, by our agreement, for he will have won his first case. So whatever happens Euathlus will be bound to pay; the court should therefore find in my favour.' But Euathlus, having learnt his lessons well, replied: 'If Protagoras wins this case, I shall not be bound to pay, for I need not pay until I win a case; but if Protagoras loses, this court will itself have decided that I need not pay; the court should therefore find for me.'

What should the court have done?

10. *Russell's Class Paradox*

Let w be the class of all those classes which are not members of themselves. Then, whatever class x may be, 'x is a w' is equivalent to 'x is not an x'. Hence, giving to x the value w, 'w is a w' is equivalent to 'w is not a w'.

Principia Mathematica, I, p. 60.

11. *The Barber*

In a certain village the barber shaves all, and only, the men who do not shave themselves. Who shaves the barber?

12. *Truth-teller Counterparts*

What I am now saying is true.

$S =_{\text{def}}$ 'S' is true.

(The sentence printed in brackets on this page is true.)

'Yields a truth when appended to its own quotation' yields a truth when appended to its own quotation.

Obey this order.

Carry out the next instruction: carry out the previous instruction.

Is 'autological' autological?

The largest number named in this book.

(The only successfully referring phrase printed in brackets on this page.)

If the instruction in Richard's paradox had read 'If the nth figure in the nth decimal is p, let the nth figure in N be p', would N have been a member of E?

'I am not going to be hanged on yonder gallows.'

Is the class of all classes that are members of themselves a member of itself?

In a certain village the barber shaves all, and only, the men who do shave themselves; is the barber bearded or clean-shaven?

13. *M-variants*

What I am now saying is either false or meaningless.

We define '*M*-heterological' as follows. An adjective *x* is *M*-heterological if used only if ' "*x*" is *x*' is either false or meaningless. Thus since ' "long" is long' is false, 'long' is *M*-heterological. Also if ' "heterological" is heterological' is meaningless, then 'heterological' is *M*-heterological. Now is '*M*-heterological' *M*-heterological or not? No answer will do, for if ' "*M*-heterological" is *M*-heterological' is true, then by the definition just given it must be either false or meaningless. But if it is false then '*M*-heterological' *is* *M*-heterological, so the statement is true, and similarly if it is meaningless it is true.

J. L. Mackie and J. J. C. Smart, *Analysis*, 13 (1953), 61–6.

14. *Prior's Family of Paradoxes*

Epimenides the Cretan says that nothing said by a Cretan is the case.

The policeman testifies that nothing the prisoner says is true, while the prisoner says that something the policeman says is true . . . the policeman cannot be right . . . the truth must be that *something* the prisoner says is true . . . either this true thing the prisoner says is the statement we know about . . . or it is something else. If it is something else, the prisoner says something else. If not . . . then the *policeman* must say something else, for the only statement of the policeman that we know about *isn't* true. So . . . if they make these two statements . . . then at least one of them must say something else besides . . . *if* neither of them says anything else, then necessarily *either* the policeman does not say that nothing the prisoner says is true *or* the prisoner does not say that something the policeman says is true. . . . Our logic does not provide any means of deciding *which* of the two statements is precluded, or of proving that *both* are.

Mr. *X*, who thinks Mr. *Y* a complete idiot, walks along a corridor with Mr. *Y* just before 6 p.m. on a certain evening, and they separate into two adjacent rooms. Mr. *X* thinks that Mr. *Y* has gone into Room 7 and himself into Room 8, but owing to some piece of absent-mindedness Mr. *Y* has in fact entered Room 6 and Mr. *X* Room 7. Alone in Room 7 just before 6, Mr. *X* thinks of Mr. *Y* in Room 7 and of Mr. *Y*'s idiocy, and at precisely 6 o'clock reflects that nothing that is thought by anyone in Room 7 at 6 o'clock is actually the case. But it has been rigorously proved, using only the most general and certain principles of logic, that under the circumstances supposed Mr. *X* just *cannot* be thinking anything of the sort.

A. N. Prior, 'On a Family of Paradoxes', *Notre Dame Journal of Formal Logic*, 2 (1961), 16–32.

15. *The Paradox of the Preface*

It is customary for authors of academic books to include in their prefaces statements such as this: 'I am indebted to . . . for their invaluable help; however, any errors which remain are my sole responsibility.' Occasionally an author will go further. Rather than say that *if* there are any mistakes *then* he is responsible for them, he will say that *there will* inevitably be some mistakes *and* he is responsible for them . . . If he has already written other books, and received corrections from readers and reviewers, he may also believe that not everything he has written in his latest book is true. His approach is eminently rational; he has learnt from experience . . . Yet [since of each of the assertions in his book he believes that it is true] he is holding logically incompatible beliefs . . . The man is being rational though inconsistent.

D. C. Makinson, *Analysis*, 25 (1965), 205–7.

But let us now face the worst thing, and as it were transfer the assertion from the preface to the body of the book . . . it can be rigorously proved, using only propositional calculus and quantification theory, that if *X* says in his book that something he says in that book is false then something that he says in his book *is* false, i.e. his statement is bound to be true . . . We cannot assert in a book that something asserted in the book is not the case, unless something other than this both is asserted in the book and is not the case. No doubt in a book in which nothing else is asserted but truths . . . anyone can inscribe . . . the *sentence* 'Something asserted in this book is false', but he cannot then *say* by this inscription . . . *that* something asserted in the book is not the case.

A. N. Prior, *Objects of Thought*, pp. 85–6.

16. *Miniac*

MINIAC: WORLD'S SMALLEST ELECTRONIC BRAIN

In this age of computers, it seems a pity that sincere but impecunious scholars should be deprived of their benefits. Herewith are presented do-it-yourself plans for constructing a computer that will answer questions not resolvable by any other present machine.

Among its many advantages, MINIAC is (1) small enough to be carried in one's watch pocket; (2) inexpensive; (3) infallible; (4) easy to build; (5) child's play to operate.

To build MINIAC:

1. Obtain a penny. (The substitution of a ha'penny will not materially affect MINIAC'S operation.)

2. Typewrite the words 'YES' and 'NO' on two pieces of paper and glue one to each side of the penny.

To operate MINIAC:

1. Hold MINIAC on the thumb and forefinger (either hand, either side up) and ask it question *A* (e.g. 'will it rain to-morrow?').

2. Flip MINIAC and allow it to come to rest.

3. Note the answer, either 'YES' or 'NO'.

Now MINIAC has given us either a true answer or a false answer. To determine which:

4. Hold MINIAC as in operating instruction 1, ask the question *B*:[1] 'Will your present answer have the same truth-value as your previous answer?' and flip.

5. Note MINIAC'S response to this question, either 'YES' or 'NO'.

Suppose MINIAC'S answer to question *B* is 'YES'. This is either a true answer or a false answer. If true, then it is true that the answer to question *A* has the same truth-value as the answer to question *B*, hence the answer to question *A* was a true answer: if false, then it is false that the answer to question *A* has the same truth-value as the answer to question *B* (which is false) hence the answer to question *A* was a true answer. In either case, if MINIAC answers 'YES' to question *B*, then its answer to question *A* was a true answer.

If MINIAC answers question *B* with 'NO', a similar line of reasoning shows that its answer to question *A* was a false answer.

The electronic character of MINIAC is obvious from the fact that there are two free electrons in the outer shell of the copper atom.

Thomas Storer, *Analysis*, 22 (1961–2), 151–2.

[1] For those who have doubts as to the legitimacy of questions of the type called 'question *B*', see (e.g.) *Introduction to Mathematical Logic*, Vol. 1, Alonzo Church, ex. 15.7, p. 105.

INDEX